T0074991

Automation in Warehouse Development

Roelof Hamberg · Jacques Verriet
Editors

Automation in Warehouse Development

Roelof Hamberg
Embedded Systems Institute
Eindhoven
The Netherlands
e-mail: roelof.hamberg@esi.nl

Jacques Verriet
Embedded Systems Institute
Eindhoven
The Netherlands
e-mail: jacques.verriet@esi.nl

ISBN 978-0-85729-967-3 e-ISBN 978-0-85729-968-0
DOI 10.1007/978-0-85729-968-0
Springer London Dordrecht Heidelberg New York

British Library Cataloguing in Publication Data
A catalogue record for this book is available from the British Library

Library of Congress Control Number: 2011938607

Cover design: eStudio Calamar S.L

Printed on acid-free paper

Springer is part of Springer Science+Business Media (www.springer.com)

Foreword

The world might be flat again,[1] but it still takes considerable time to ship goods from one side to the other. Distribution centres take up a central role in the supply chain to stock goods coming from a multitude of producers and ship customer-specific orders to the end client. Increasingly high demands have to be fulfilled in a cost-efficient manner.

In 2006, we took up the challenge to tackle "one of the last frontiers" in industrial automation: unstructured item picking by human order pickers in distribution centres. At the time, progress in vision technology and robotics gave us the impression this was an achievable goal. However, automation of the "hardware" is only half the problem, and not even the most difficult half.

A distribution centre fulfils many functions and its design is based on many years of experience. Item picking by human operators takes up a central role, and therefore design rules incorporate aspects of the whole system. You simply cannot replace one aspect, in isolation, by an "automated" version without considering the total system. That was the real challenge we put into this project: how to incorporate a new and yet unknown function in an optimal way for the system as a whole.

And what a challenge it was.

The "industry-as-laboratory" philosophy of the Embedded Systems Institute was enthusiastically embraced by research groups of the universities of Eindhoven, Delft, Twente, and Utrecht. Right from the start these groups faced the challenging ("frustrating", some would say) deadlock situation. Groups working on component design needed overall system aspects of the yet to be developed system, groups working on system design needed details of the yet to be developed components. And both were not able to provide the other with the requested information.

[1] See Thomas Friedman's *The World is Flat: A Brief History of the Twenty-First Century* published by Farrar, Straus and Giroux.

Now, 5 years down the line, this book summarises what we have achieved. One of our main lessons is that there is no such thing as a "one solution suits all". We will have to remain flexible in applying the lessons learned in new situations. If they do not fit, use other ones or develop new solutions if others do not exist. That was what this project was really about: Flexible Automated Logistics CONcepts (FALCON).

Veghel, April 2011

Gert Bossink
Director R&D
Vanderlande Industries

Preface

This book marks the end of the Dutch BSIK Falcon project, addressing the model-driven development of automated logistic systems. In particular, the warehouse and distribution systems as developed by Vanderlande Industries, a leading supplier of integrated logistics systems for automation of warehouses, were selected as Falcon's carrying industrial case.

This books is the sixth in a series of books, all reporting on the large 5-year industry-as-laboratory projects, as executed by the Embedded Systems Institute (ESI), in close collaboration with its industrial and academic partners.[2]

The Falcon project was executed from October 2006 until September 2011. It was carried out by a consortium consisting of the Embedded Systems Institute, Vanderlande Industries, Demcon Advanced Mechatronics, Delft University of Technology, Eindhoven University of Technology, Eurandom, the University of Twente, and Utrecht University, and lead by ESI, together encompassing 95 FTE.

Falcon focuses on model-driven development within the context of warehouse automation with a specific emphasis on warehouse control and enabling technologies for the automation of warehouse functions. Model-driven development provides a means to handle increased system complexity by focusing on the problem domain instead of the solution domain. It is a development approach that focuses on creating models and system abstractions, as a means to increase productivity and quality by simplifying design processes and supporting design decisions, whilst promoting communication between all parties involved.

The Embedded Systems Institute has, in its previous projects, already shown the benefits of model-driven development in several other domains. The Falcon project adds a new domain by successfully applying this approach to the design of warehouses, robotic warehouse components, and warehouse management and control systems. Key results of the project include highly modular reference architectures that support the model-driven development of warehouse management and control

[2] The books of the earlier industry-as-laboratory projects Boderc, Tangram, Ideals, Trader, and Darwin are available on ESI's website: http://www.esi.nl/publications/books/.

systems; simple aspect models that effectively guide the development of new warehouse concepts or the configuration of new warehouse systems; a versatile automated item-picking workstation integrating a novel underactuated robot hand and new item recognition and localisation algorithms; and a scalable and robust shuttle-based transportation system with similar performance as conventional conveyor-based transportation systems.

I would like to thank the participants of the Falcon project for their commitment and contributions to the project: together they have secured Falcon's success. The financial support by Vanderlande Industries and the Dutch Ministry of Economic Affairs (through AgentschapNL) are gratefully acknowledged. I would also like to thank Springer for their willingness to publish this book, with which the Embedded Systems Institute wishes to share the most important results of the Falcon project with a broad audience, in both industry and academia.

Eindhoven, April 2011

Prof. Dr. Ir. Boudewijn Haverkort
Scientific Director and Chair
Embedded Systems Institute

Acknowledgments

Automation in warehouse development is a result of the Falcon project conducted under the responsibility of the Embedded Systems Institute with Vanderlande Industries as the carrying industrial partner. This project is partially supported by the Netherlands Ministry of Economic Affairs under the Embedded Systems Institute (BSIK03021) program.

We gratefully acknowledge the cooperation with the employees of Vanderlande Industries throughout the 5-year project. Their warehouse domain knowledge and our partners' applied research have formed the basis for *Automation in warehouse development*, a book targeted at academic researchers and industrial practitioners describing state-of-the-art research on warehouse automation and model-based warehouse design.

We would like to thank the employees of our partners (Embedded Systems Institute, Vanderlande Industries, Demcon Advanced Mechatronics, Delft University of Technology, Eindhoven University of Technology, University of Twente, Utrecht University, and Eurandom) for writing the chapters of this book. We would also like to thank the reviewers from Vanderlande Industries, the Embedded Systems Institute, and our academic partners. Their constructive feedback has improved the quality of the book.

Contents

Part I
Introduction

Chapter 1
The Falcon Project: Model-Based Design of Automated Warehouses

Roelof Hamberg and Jacques Verriet

Abstract Due to the increasing scarcity of human operators and the fast development of technology, there is a trend towards higher levels of automation in warehouses. The increased automation level results in a large warehouse complexity and a correspondingly large warehouse design effort. To support the warehouse designer in his effort, Vanderlande Industries and the Embedded Systems Institute started the Falcon project in October 2006. The project focuses on the development of professional systems using a model-driven approach. Vanderlande Industries is the project's carrying industrial partner, which means that their automated warehouse development is used as the basis for the project. This chapter introduces the partners and the research themes of the Falcon project and describes the structure of the book, which presents the results of the Falcon project.

1.1 Introduction

Someone shopping in a supermarket probably does not realise how the supermarket's products ended up at the shelves from which they were taken and placed in his shopping basket. *Warehouses*, or *distribution centres*, play an essential role here; they are responsible for the effective distribution of goods from suppliers to customers. A typical retail warehouse decouples the producer of a product from its end customers: it has many producers as its suppliers and many supermarkets as its customers. This greatly reduces transportation cost and time, because the transportation effort is shared over many producers and consumers.

R. Hamberg · J. Verriet (✉)
Embedded System Institute, P.O. Box 513, 5600 MB Eindhoven, The Netherlands
e-mail: jacques.verriet@esi.nl

R. Hamberg
e-mail: roelof.hamberg@esi.nl

R. Hamberg and J. Verriet (eds.), *Automation in Warehouse Development*,
DOI: 10.1007/978-0-85729-968-0_1, © Springer-Verlag London Limited 2012

A (retail) warehouse can not only save transportation cost of its supermarket customers outside the supermarkets, it can even save handling costs inside the supermarkets. This requires a customer-specific delivery of ordered goods: a warehouse has to deliver goods in such a way that the supermarkets' shelves can be replenished efficiently. This *value-added service*, called *store-friendly delivery*, requires that products of one supermarket shelf are packed in the same delivery container and that these containers are arranged in a specific sequence. This packing and sequencing greatly reduces the travel time of the supermarkets' replenishment personnel, but requires a more controlled and thus more complex flow of goods in a warehouse.

1.2 Warehouses

A warehouse, or distribution centre, is a facility that stores products from many different suppliers for further distribution to their customers. Generally, five main processes can be identified in a warehouse: receiving, storage, picking, consolidation, and shipping [19]. The *receiving* process involves the arrival of products from a supplier and may include a preparation of storage by repacking of the products into special storage containers, e.g. product totes. The *storage* process is responsible for placing received products into storage areas; there the products remain until they are needed. The *picking* process is triggered by the reception of customer orders. It involves the retrieval of products from storage and placing them in special order containers, e.g. order totes. These order containers are collected in the *consolidation* process and combined in shipping containers, e.g. pallets. The *shipping* process takes care of the shipment of the combined order containers to the customers. The areas implementing these warehouse processes are connected by a *transportation* system, e.g. a conveyor system.

Figure 1.1 shows a sketch of a retail warehouse in the United Kingdom. Figure 1.2 shows pictures of its receiving, storage, picking, consolidation, and shipping areas.

1.2.1 Trends

It is not uncommon for a modern warehouse to provide storage for tens of thousands of products from thousands of different suppliers. Moreover, modern warehouses often have to deliver these products to a large variety of different customers. These customers may include other warehouses and various types of stores, but also internet customers. Each of these customers has its own delivery requirements, for instance, the store-friendly delivery to supermarkets.

Value-added services, like store-friendly delivery, is just one of the trends in (retail) warehousing: warehouses tend to serve a growing variety of customers. For instance, several retailers are introducing new store formats to address new markets [20]. Each (type of) customer has its own specific delivery requirements. Other

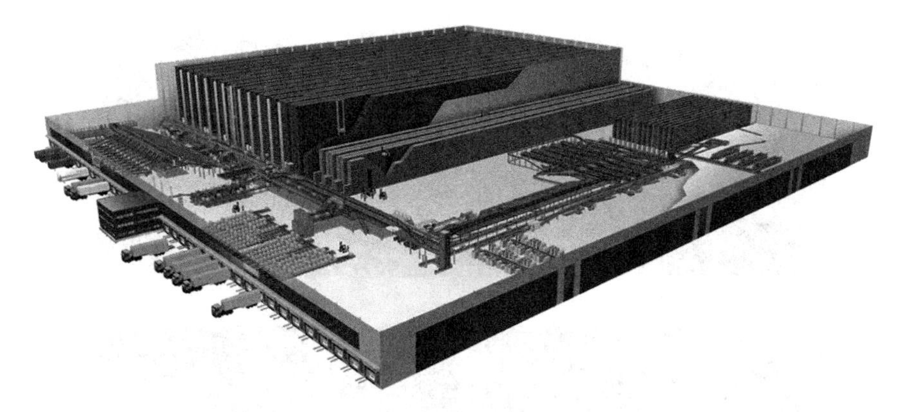

Fig. 1.1 Schematic overview of a retail warehouse. The size of the trucks on the left indicate the size of the warehouse

warehousing trends include increasing delivery frequencies and decreasing order sizes. Because of e-commerce and the emergence of smaller convenience stores, warehouses are required to deliver smaller orders to their customers. Moreover, warehouse responsiveness is increasing, as its customers require orders to be delivered the next day [20].

Another important warehousing trend, especially in Western Europe and North America, is a higher level of automation. This trend is triggered by a high pressure to decrease the operational costs [27]. Another trigger of the automation trend is the cost and availability of resources. For instance, there is a scarcity of suitable warehouse space and qualified staff is hard to attract [1]. Traditionally, warehouse operations, particularly order picking, are performed by human operators travelling through the warehouse. Traditional warehouses are built around these human operators, because they account for more than half of a warehouse's operational cost [6]. However, the availability of human operators is a growing concern. The required workforce is increasingly harder to obtain due to issues relating to health and safety, night-time working hours, hard and uninspiring work, and low wages. High labour mobility leads to additional problems with respect to training and quality assurance. Thus, the automation of warehouse operations seems the only answer to the cost and unavailability of human operators.

1.2.2 System Development

Designing a warehouse is a very challenging task and the trends identified in the previous section only increase this challenge. Warehouse design involves a large number of decisions to be made. Three levels of warehouse design can be distinguished [19]: strategic level, tactical level, and operational level. These levels address long-term, medium-term, and short-term decisions to be made. A comprehensive overview of

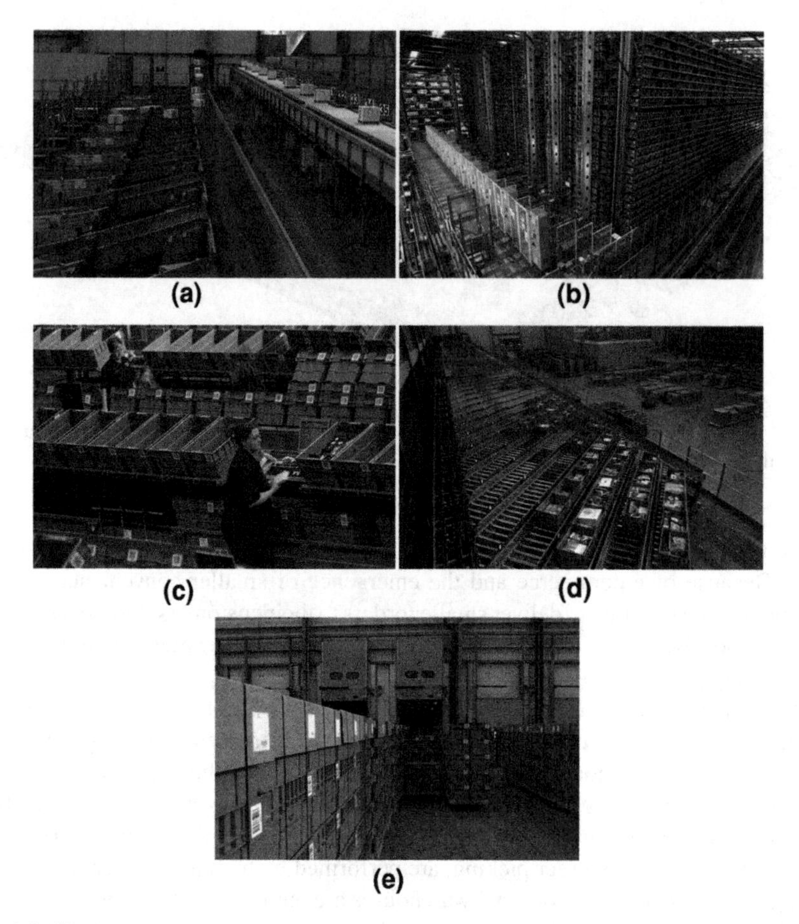

Fig. 1.2 Warehouse process areas: **a** receiving, **b** storage, **c** picking, **d** consolidation, **e** shipping

warehouse design models is given by Gu et al. [9]. They identify five main types of decisions to be made at the strategic and tactical level.

1. *Overall structure* determines the functional areas of a warehouse and defines how orders are assembled.
2. *Sizing and dimensioning* identifies the capacities of the warehouse's functional areas.
3. *Layout* defines the physical arrangement of the warehouse areas.
4. *Equipment selection* specifies the material handling equipment to be deployed.
5. *Operation strategy* identifies the high-level operational strategies.

At the operational level, decisions have to be made regarding the execution of the processes defined at the strategic and tactical level. This includes *planning*, i.e. assigning tasks to resources, and *scheduling*, i.e. sequencing assigned tasks. Many of these operational decisions are addressed in the review by Gu et al. [8].

Obviously, the decisions made at one level are not independent, neither are decisions made at different levels. Moreover, there is not a single warehouse, which meets all requirements. This makes warehouse design a very challenging task. Although there are many models for aspects of warehouse design, little integrated warehouse design support is available [19].

1.3 The Falcon Project

In October 2006, the Falcon project was started. Falcon is an acronym for Flexible Automated Logistics Concepts. The goal of the Falcon project is to bridge the gap between component design and system design, thereby filling the lack of integrated warehouse design support identified by Rouwenhorst et al. [19].

Falcon's project plan states its purpose as "the development of techniques and tools for the design and implementation of professional systems, with a special focus on the optimisation and decomposition of global requirements on system performance, reliability, and cost using a model-driven approach". As application domain for this project, the field of warehouses and distribution centres was selected with a special focus on the automation of order picking. Falcon's main project goal has been decomposed into three objectives:

- The development of tools and methods for the design of warehouses;
- The development of control methods to make a warehouse system perform as intended; and
- The development of robotic item-picking solutions.

Each of the goals was to be demonstrated using prototypes and proofs-of-concept within an industrial context, i.e. the warehouse and distribution domain.

The Falcon project has been carried out according to the *industry-as-laboratory* philosophy [18]. In an industry-as-laboratory project, researchers have a close cooperation with industrial practitioners and address industrial problems using state-of-the-art academic research. The Embedded Systems Institute has successfully applied this research paradigm, with different so-called carrying industrial partners, in a series of projects: Boderc [10], Tangram [22], Ideals [25], Trader [16], and Darwin [24].

The Falcon project involved a collaboration of several partners, both academic and industrial:

- Embedded Systems Institute;
- Vanderlande Industries B.V.;
- Demcon Advanced Mechatronics;
- Delft University of Technology;
- Eindhoven University of Technology;
- University of Twente;
- Utrecht University; and
- Eurandom.

Fig. 1.3 Falcon project team

Each partner had a different role in the project. The Embedded Systems Institute's role included industrial research, project management, overall technical supervision, and consolidation of project results. The role of Vanderlande Industries is that of carrying industrial partner, which entails defining the context of the project and providing access to industrial experts and knowledge needed to reach the industrial goals. The role of the academic partners has been the research into the industrial problems of Vanderlande Industries. Demcon's role involved the integration of some of the academic research results into an industrial prototype.

Figure 1.3 shows a picture of the Falcon project team during the Falcon research demonstration day in July 2008 in Vanderlande Industries' Innovation Centre in Veghel.

1.4 Book Outline

This book describes the research results of the Falcon project validated against Vanderlande Industries' industrial goals. The book title "Automation in warehouse development" deliberately allows a double interpretation, as two types of automation are addressed. On the one hand, the book covers automation of warehouse functionality to address the cost and unavailability of human warehouse operators. On the other hand, automation of the warehouse development process is addressed by applying a model-based design approach. This kind of automation is intended to support warehouse designers in their challenging task of designing effective (automated) warehouses.

The Falcon project results have been grouped into four main research themes:

- Decentralised control engineering;
- Models in system design;

- Automated item handling; and
- Transport by roaming vehicles.

These themes define the structure of the book: the book has four main parts, each covering one of the research themes. The parts and their constituting chapters are described in the following subsections.

1.4.1 Decentralised Control Engineering

One of the main challenges in designing a warehouse is the design of an effective warehouse management and control system to control the warehouse operations. Traditionally, these systems are centralised systems optimised for a warehouse's business process. Their optimised nature makes it difficult to adapt them in case of business changes or to reuse their functionality for other warehouses. This has recently been identified in the warehousing domain: decentralised controllers have been shown to be more flexible than their centralised counterparts [7, 11]. Similar results have been shown in domains similar to the warehousing domain, the manufacturing domain [12, 17] in particular. In the latter domain, this has led to the definition of reference architectures, e.g. PROSA [23] and ADACOR [13], that facilitate the design of decentralised manufacturing control systems.

Decentralised control engineering is the subject of Part II, which consists of Chaps. 2, 3, and 4. It describes the development of methods for the design of decentralised warehouse management and control systems. These methods are based on reference architectures consisting of generic components from which specific warehouse management and control systems can be constructed.

Two reference architectures are presented. Chapter 2 introduces a hierarchical reference architecture consisting of two types of components that can be configured using a collection of structural parameters. These parameterised components interact using standardised interaction protocols, which can be configured using behavioural parameters, i.e. *business rules*, to obtain the desired warehouse management and control system behaviour. This goes a step beyond the existing reference architectures for decentralised manufacturing control as they incorporate behaviour using roles.

Chapter 3, considers another type of organisation than the hierarchies of Chap. 2. It describes a reference architecture based on organisations of autonomous components called *agents*: organisations of agents enact functional roles of the system. Together the enacted roles provide the desired system behaviour.

The reference architectures presented in Chaps. 2 and 3 can form the basis for a more structured way of developing warehouse management and control systems. This so-called *model-driven software engineering approach* is discussed in Chap. 4: it describes the benefits of model-driven development of warehouse management and control systems over traditional development methods.

1.4.2 Models in System Design

As we identified in Sect. 1.2.2, many decisions have to be made when designing a warehouse system. These decisions are often made based on the experience of the warehouse designer. A comprehensive systematic method for designing warehouses is not readily available in literature [2]. Simulation is a common tool used to help the warehouse designer. Unfortunately, simulation can only be applied in the later stages of the warehouse design to validate design decisions made earlier. The same holds for many of the models and tools presented in literature [8]. Except simple rules-of-thumb, like Little's law [14], warehouse designers have few means that aid them during the early stages of the development process.

The models in system design theme addresses the lack of analysis models usable during the early phases of warehouse design. This subject is addressed in Part III, which consists of Chaps. 5, 6, and 7 of the book. Chapter 5 describes a method, called *effective process time*, which is able to capture the essential parameters of warehouse components and aggregate them to obtain simpler component models. Chapter 5 also shows how these simplified component models can be used to build hierarchical warehouse performance models that can be used to guide the warehouse design process.

Chapter 6 addresses the application of models in the development of new warehousing concepts. It describes models for a highly automated warehouse concept: an automated case picking concept. The basic principles of the high-level method of the Boderc project [10] are brought into practice. With this industrial case, ample evidence is gathered to claim that partial system modelling focusing on critical issues has a positive cost/benefit ratio with regard to its role in system development.

Chapter 7 shows some examples of simple models that allow a warehouse designer to make decisions in the early phases of warehouse system configuration. It presents two lightweight simulation tools that abstract from the details of the material handling system and validates these tools by applying them to a goods-to-man warehouse concept. The central idea is that a full warehouse layout is not needed in order to build performance models that are sufficiently accurate to assist in making warehouse configuration decisions.

1.4.3 Automated Item Handling

In Sect. 1.2.1, we have identified the trend towards higher levels of automation. This trend has been addressed in the Komrob project [22], which has focused on the automated handling of *cases*, large box-shaped products that can be transported individually. In Sect. 1.2.1, we have also identified the trend towards smaller orders and higher order frequencies. This increases a warehouse's need to deliver individual *items* instead of cases. Items are small products that cannot be transported individually. Because of the large variety in items' shape, size, packaging, colour, and texture, item picking is a task that is hard to automate. Thus nearly all item picking is performed by human operators.

Part IV, consisting of Chaps. 8 through 12, covers automated item handling. It addresses the automation of the challenging item-picking task. It shows how the recent development of effective object recognition and localisation algorithms [3, 15] and the concept of underactuation [4] can be applied to automate item handling.

Chapter 8 addresses the capability of a robotic item-picking workstation built from commercial-of-the-shelf components: an industrial robot, standard vision components, and a suction gripper. The results show that most items can be handled by a commercial-of-the-shelf solution, but also that a significant range of items cannot be handled.

Chapters 9 through 12 take the commercial-of-the-shelf solution as a benchmark. They present more advanced gripping and vision solutions to overcome the limitations of a commercial-of-the-shelf solution.

Chapter 9 describes a model-based approach to design a robust and flexible gripper from the characteristics of the items to be grasped. The robustness and flexibility of this hand come from its *underactuated* and *compliant* nature: it has a minimum number of electronic components, i.e. one motor and no sensors, while its resulting grasp is adaptive with respect to different item shapes.

Chapter 10 presents a method to learn items to be picked in a warehouse by constructing a database containing the items' most distinctive features and its grasping points. The algorithms presented in Chap. 10 only consider 2D features. This is not sufficient for the recognition of featureless items, e.g. white boxes. This is addressed in Chap. 11, which adds 3D features to the item database. Chapters 10 and 11 both explain how the item database can be used to recognise items and localise the best items for gripping.

The solutions presented in Chaps. 9, 10, and 11 have been integrated in an industrial prototype. The corresponding integration process and the comparison of the resulting prototype and the commercial-of-the-shelf solution presented in Chap. 8 are discussed in Chap. 12.

1.4.4 Transport by Roaming Vehicles

Transportation is a warehouse task, which is commonly automated. The traditional way to automate transportation involves the usage of conveyors. These components have a good performance and a high availability. However, they often constitute a single point of failure: the breakdown of a single conveyor may bring the entire warehouse to a standstill, because redundancy in the transportation system is generally too expensive. Recently, several vehicle-based transportation systems have been introduced [5, 26]. These solutions are characterised by the confinement of the vehicle to (physical or marker-based) rails.

Part V, which comprises Chaps. 13 and 14, addresses the central challenges for a transportation system consisting of (autonomous) roaming vehicles that are not restricted to travel a defined infrastructure. A prerequisite of collision-free vehicle motion is the vehicles' awareness of their location in the warehouse. This is addressed in Chap. 13 which describes techniques for a vehicle to build a map of its environment and simultaneously use this map to determine its own position using natural features instead of special markers.

Chapter 14 presents techniques to coordinate the movements of a group of autonomous vehicles for the fulfilment of transportation tasks. This chapter presents an architecture in which the vehicle coordination can be implemented using high-level or low-level coordination algorithms. It evaluates these coordination algorithms using the conveyor-based transportation system of an existing retail warehouse as a reference.

1.4.5 Reflections and Appendices

Besides the four parts covering Falcon's research themes, this book contains a reflection chapter and two appendices. Chapter 15 reflects on the Falcon project. It discusses the main project results as well as its organisational aspects. It also describes the impact of the project on its industrial partners and provides recommendations for industry-as-laboratory projects.

Appendix A provides address information of the Falcon partners. Appendix B gives an overview of the publications of the project.

References

1. Angel BWF, van Damme DA, Ivanovskaia A, Lenders RJM, Veldhuijzen RS (2006) Warehousing space in Europe: meeting tomorrow's demand. Capgemini Consulting Services, Utrecht
2. Baker P, Canessa M (2009) Warehouse design: A structured approach. Eur J Oper Res 193: 425–436
3. Bay H, Ess A, Tuytelaars T, Van Gool L (2008) Speeded-up robust features (SURF). Comput Vis Image Underst 110:346–359
4. Birglen L, Laliberté T, Gosselin C (2008) Underactuated robotic hands. Springer tracts in advanced robotics, vol 40. Springer, Berlin
5. Fraunhofer-Institut für Materialfluss und Logistik (2011) MultiShuttle Move. http://www.iml.fraunhofer.de/en/fields_of_activity/autonomous_transport_systems/multishuttle_move.html, Viewed May 2011
6. Frazelle E (2002) World-class warehousing and material handling. McGraw-Hill, Columbus
7. Graves RJ, Wan VK, van der Velden J, van Wijngaarden B (2008) Control of complex integrated automated systems—system retro-fit with agent-based technologies and industrial case experiences. In: Proceedings of the 10th international material handling research colloquium
8. Gu J, Goetschalckx M, McGinnis LF (2007) Research on warehouse operation: A comprehensive review. Eur J Oper Res 177:1–21

9. Gu J, Goetschalckx M, McGinnis LF (2010) Research on warehouse design and performance evaluation: A comprehensive review. Eur J Oper Res 203:539–549
10. Heemels M, Muller G (eds) (2006) Boderc: model-based design of high-tech systems. Embedded Systems Institute, Eindhoven
11. Kim BI, Graves RJ, Heragu SS, St. Onge A (2002) Intelligent agent modeling of an industrial warehousing problem. IIE Trans 34:601–612
12. Leitão P (2009) Agent-based distributed manufacturing control: a state-of-the-art survey. Eng Appl Artif Intel 22:979–991
13. Leitão P, Restivo F (2006) ADACOR: A holonic architecture for agile and adaptive manufacturing control. Comput Ind 57:121–130
14. Little JDC (1961) A proof for the queuing formula: $L = \lambda W$. Oper Res 9:383–387
15. Lowe DG (2004) Distinctive image features from scale-invariant keypoints. Int J Comput Vis 60:91–110
16. Mathijssen R (ed) (2007) Trader: Reliability of high-volume consumer products. Embedded Systems Institute, Eindhoven
17. Monostori L, Váncza J, Kumara SRT (2006) Agent-based systems for manufacturing. Ann CIRP 55:697–720
18. Potts C (1993) Software-engineering research revisited. IEEE Softw 10:19–28
19. Rouwenhorst B, Reuter B, Stockrahm V, van Houten GJ, Mantel RJ, Zijm WHM (2000) Warehouse design and control: Framework and literature review. Europ J Oper Res 122: 515–533
20. Sadler J (2006) The future of European food and drinks retailing: Implications of fast growth formats and private label for brand manufacturers. Business Insights Ltd, London
21. Schraft RD, Westkämper E (eds) (2005) Roboter in der Intralogistik–aktuelle Trends, moderne Technologien, neue Anwendungen. Fraunhofer-Institut für Produktionstechnik und Automatisierung, Stuttgart
22. Tretmans J (ed) (2007) Tangram: Model-based integration and testing of complex high-tech systems. Embedded Systems Institute, Eindhoven
23. Van Brussel H, Wyns J, Valckenaers P, Bongaerts L, Peeters P (1998) Reference architecture for holonic manufacturing systems: PROSA. Comput Ind 37:255–274
24. van de Laar P, Punter T (eds) (2011) Views on evolvability of embedded systems. Springer, Dordrecht
25. van Engelen R, Voeten J (eds) (2007) Ideals: Evolvability of software-intensive high-tech systems. Embedded Systems Institute, Eindhoven
26. Wurman PR, D'Andrea R, Mountz M (2008) Coordinating hundreds of cooperative, autonomous vehicles in warehouses. AI Mag 29:9–20
27. Wyland B (2008) Warehouse Automation: How to implement tomorrow's order fulfillment system today. Aberdeen Group, Boston

Part II
Decentralised Control Engineering

Chapter 2
A Reference Architecture Capturing Structure and Behaviour of Warehouse Control

Jacques Verriet and Bruno van Wijngaarden

Abstract Warehouse management and control systems are responsible for the operations in a warehouse. These systems are usually very complex due to the delivery requirements of the warehouse customers. These delivery requirements are often very specific for a group of customers, which makes it hard to reuse warehouse management and control functionality for other warehouses. In this chapter, we will introduce a method that will greatly increase the reusability of warehouse management and control functionality. We present a reference architecture that supports the development of warehouse management and control systems. This modular reference architecture is based on functional components, which can be configured using structural and behavioural parameters. This configuration is sufficient to build a warehouse management and control system. This chapter introduces the reference architecture and demonstrates how it can be used to decrease the warehouse management and control system development effort.

2.1 Introduction

The operations in a warehouse are controlled by a warehouse management and control system (WMCS). To achieve a high warehouse performance, a WMCS needs to use a warehouse's scarce resources in an efficient manner. A WMCS is commonly seen as a layered system. For instance, Ten Hompel and Schmidt [8] distinguish several layers of WMCS control. These include strategic control by an enterprise resource planning (ERP) system, functional control by a warehouse management

J. Verriet (✉)
Embedded System Institute, P.O. Box 513, 5600 MB Eindhoven, The Netherlands
e-mail: jacques.verriet@esi.nl

B. van Wijngaarden
Vanderlande Industries B.V, Vanderlandelaan 2, 5466 RB, Veghel, The Netherlands
e-mail: bruno.van.wijngaarden@vanderlande.com

R. Hamberg and J. Verriet (eds.), *Automation in Warehouse Development*,
DOI: 10.1007/978-0-85729-968-0_2,

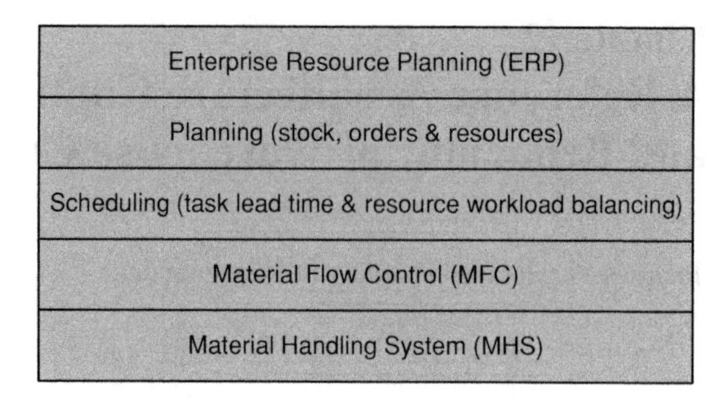

Fig. 2.1 Layers of warehouse functionality

system (WMS), section control by a warehouse control system (WCS), and process control by a material flow controller (MFC).

In this chapter, we will distinguish five layers of WMCS functionality, which are illustrated by Fig. 2.1. The top layer is the ERP system, which is responsible for the high-level management of orders and stock. The second layer, the planning layer, handles the assignment of orders to stock and equipment resources. The planning layer is also responsible for keeping sufficient stock levels by replenishing low stock levels and relocating superfluous stock. The scheduling layer is responsible for balancing the system to obtain the optimal system performance. The scheduling layer's responsibilities include the prioritisation of the tasks that have been assigned to resources. Scheduling selects tasks and forwards them to the MFC layer, which controls the warehouse equipment and provides equipment status information to the scheduling layer. The MFC layer provides an interface to the bottom layer, the material handling system (MHS), which contains the warehouse equipment handling the actual execution of warehouse tasks.

Because of the size and complexity of modern warehouses, the development of a WMCS is very difficult and time consuming. As identified in Chap. 1, an important element of the complexity of a warehouse involves the product and delivery requirements of the warehouse customers. Another part of the complexity of WMCS development is due to the lack of WMCS development support. The layered architecture shown in Fig. 2.1 provides some design guidelines, but it does not provide sufficient support for making the design decisions needed to develop an effective WMCS. This lack of support increases the risk of developing a customer-specific WMCS, whose functionality cannot be reused for other warehouses.

In this chapter, we will address the lack of WMCS development support. We hypothesise that a WMCS reference architecture can reduce the WMCS development effort and allows WMCS functionality to be reused for different warehouses. We will test this hypothesis by defining a WMCS reference architecture that captures both the structure and the behaviour of warehouse control. We will illustrate how our WMCS reference architecture decreases the WMCS development effort and increases the reusability of WMCS functionality.

2.1.1 Decentralised Warehouse Control

Since modularity and loose coupling are important characteristics of systems built from reusable components, we let ourselves be inspired by the recent research into agent-based control systems. Such systems consist of a collection of autonomous agents, each having a limited scope. Examples from the warehousing domain include the work by Kim et al. [4] and Graves et al. [3]. They study an agent-based WMCS for warehouses with a man-to-goods and a goods-to-man picking concept, respectively. They propose a hierarchical agent-based WMCS, in which high-level agents compute a global schedule. The low-level agents constantly reschedule their work to match the real-time conditions. If they find a better schedule, they negotiate with the high-level agents for a schedule change. The high-level agents will allow the proposed schedule change if the new schedule does not deteriorate the global schedule. The results of Kim et al. [4] and Graves et al. [3] show that the constant rescheduling by the low-level agents has a positive influence on the system performance. Although the WMCS agents do not have full system knowledge, they have a similar performance as the original WMCS under normal circumstances and a better performance in case of exceptions.

Decentralised control has also been applied in domains with similarities to the warehousing domain. Other examples of decentralised control mainly involve the manufacturing domain. There are many publications describing the benefits of decentralised manufacturing systems. An example is the engine manufacturing line described by Fleetwood et al. [2], who present a decentralised control system that has a better performance and flexibility than its centralised counterpart.

Many applications of decentralised control systems are very specific for the application. However, a few reference architectures have been proposed for decentralised control systems. Examples are PROSA [9] and ADACOR [5]. The applicability of these reference architectures has been demonstrated using a variety of applications in the manufacturing domain. These reference architectures improve reusability by identifying generic roles and corresponding interaction protocols. Within the Falcon project, Moneva et al. [7] defined a WMCS reference architecture, which standardises WMCS functionality using roles and interaction protocols similar to PROSA's. These generic roles and interaction protocols allow warehouse operations to be performed by a collection of agents organised in a hierarchy matching the underlying material handling system.

2.1.2 Outline

In this chapter, we will go one step further in standardising WMCS functionality than Moneva et al. [7]. We will present a WMCS reference architecture that standardises not just the components and their interfaces, but also the components' functionalities. This is achieved by using generic behaviours with local business rule plug-ins, which can be used to make a behaviour application specific. This standardisation allows a

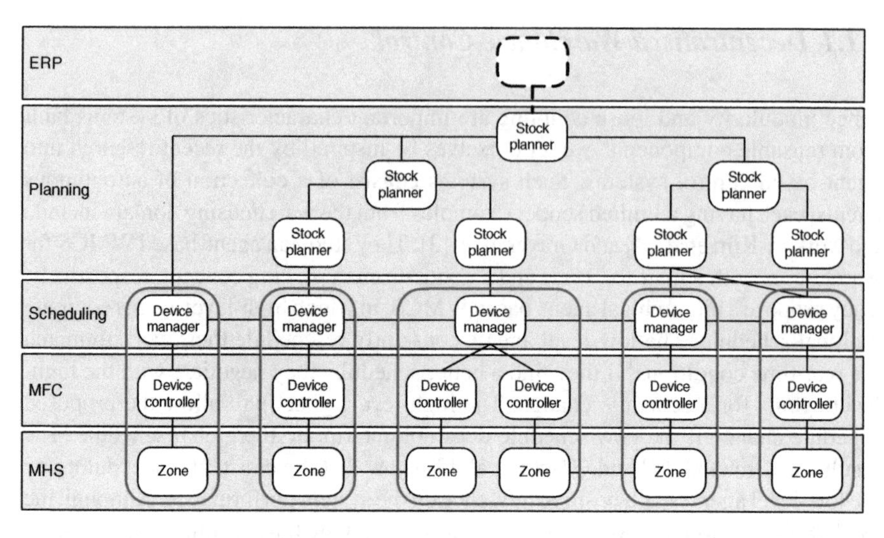

Fig. 2.2 WMCS reference architecture

WMCS to be generated automatically by identifying the components, their interconnections and behaviours, and by specifying the business rules of these behaviours.

The chapter is organised as follows. Section 2.2 describes the structural and behavioural components of the WMCS reference architecture. Section 2.3 introduces an automated case picking system, which will be used to illustrate the benefits of our reference architecture. A prototype implementation of the reference architecture is described in Sect. 2.4. Sections 2.5 and 2.6 present the initial results validating the new architecture. Section 2.7 describes the current status of the reference architecture and an outlook into the future.

2.2 Warehouse Management and Control Reference Architecture

In this section, the WMCS reference architecture is explained in terms of its structural and behavioural components.

2.2.1 WMCS Components

Our decentralised WMCS reference architecture is a layered system, which distinguishes the same layers as Fig. 2.1. The layers of the architecture and their components are shown in Fig. 2.2. The WMCS reference architecture distinguishes four types of components each corresponding to a different layer of the WMCS hierarchy: stock planners, device managers, device controllers, and (material handling) zones.

The *stock planners* reside in the planning layer. The stock planners' main responsibility is the delivery of goods to its (local) customers. This delivery service is determined by its connected device managers, which define which types of goods can be delivered from a stock planner's local stock. Figure 2.2 shows that the stock planners form a tree. This tree structure allows a logical clustering of stock planners, for instance based on the type of goods they can deliver.

The *device managers* in the scheduling layer couple the abstract stock planners in the planning layer to the concrete device controllers in the MFC layer. They are responsible for sequencing the tasks assigned by the stock planners. Upon the assignment or completion of a task, a device manager selects the next task to be executed and forwards it to its underlying device controller. Figure 2.2 shows that the device managers form a network of device managers. This network is an abstraction of the warehouse topology in the MHS layer. This abstraction is used to determine how work is to be handed over in the scheduling layer in order to obtain the desired flows of goods through the warehouse. The connections in this network can be seen as producer-consumer relationships between device managers.

The *device controllers* in the MFC layer are responsible for the coordination of task execution by the material handling zones. Figure 2.2 shows that there is a one-to-one correspondence between zones and device controllers. It also shows that each device controller is connected to one device manager, but that a device manager can be connected to several device controllers. Then the device manager is responsible for dividing the work over the underlying device controllers.

The *zones* in the MHS layer are responsible for the actual execution of warehouse tasks. Examples of material handling zones are miniloads, order-picking workstations, and transport loops. Figure 2.2 shows connections between the zones; these correspond to the physical connections between the warehouse equipment.

Figure 2.2 shows a similarity to the organisation-based WMCSs in Chap. 3: a device manager and its underlying device controllers and zones can be seen as the presences of a single MASQ agent in the scheduling, MFC, and MHS communication spaces. This has been illustrated by the rounded rectangles in Fig. 2.2.

Both the zones and the device controllers are equipment-specific, and therefore reusable, components. In our WMCS reference architecture, the stock planner and device managers are generic components. Both types of components have parameters from which a structure as shown in Fig. 2.2 can be constructed. The stock planners' parameters include their parent and children in the tree of stock planners and the connected device managers. The device managers' parameters include the neighbours in the network of device managers and the capabilities of the underlying device controllers.

2.2.2 WMCS Behaviours

In Sect. 2.1, we have defined the components of our WMCS reference architecture. A collection of these components needs to behave in such a way that the underlying

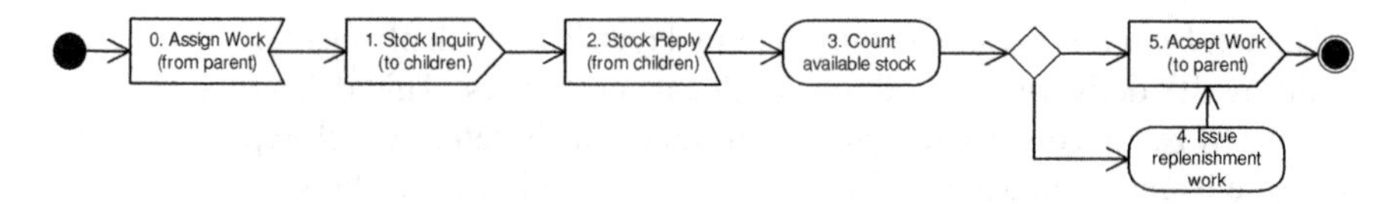

Fig. 2.3 AssignWork behaviour activity diagram

warehouse equipment operates in an efficient manner. The main challenge in designing an effective WMCS is designing the behaviours of its components that together provide the desired system behaviour. We will focus on the behaviour of the stock planners and device managers, as their high-level behaviour has the largest influence on system behaviour.

To alleviate the system design effort, we have standardised component behaviour. We have defined a collection of generic behaviours for stock planners and device managers: their interaction has been fixed with respect to the interacting agents and the interface object types. Each of these behaviours is triggered by the reception of messages matching generic interfaces. In this section, we will describe some examples of our reference architecture's behaviours implementing the generic interaction protocols; this includes the object types being communicated.

2.2.2.1 Delivery Behaviour

The main goal of the stock planners is the delivery of goods to its customer stock planners. Several generic behaviours have been defined to control this delivery. An example is the AssignWork behaviour shown as a UML activity diagram in Fig. 2.3. This behaviour is triggered by the reception of an AssignWork message; this is a standardised interface specifying the goods to be delivered and the container type in which to deliver them. After receiving the message, the AssignWork behaviour sends a StockInquiry message to its child stock planners to ask for the availability of the requested goods. It then waits for the corresponding StockReply messages and computes the available stock level. If the stock level does not suffice, the behaviour issues replenishment work. This involves sending and receiving several messages, but this has not been detailed in Fig. 2.3. The AssignWork behaviour ends by sending an AcceptWork message to acknowledge the original AssignWork message.

The completion of the AssignWork behaviour will asynchronously trigger another delivery behaviour, the ForwardWork behaviour, which is shown in Fig. 2.4. This behaviour starts by checking whether all stock needed for a work assignment is available in its children's local stock. If sufficient stock is available, it will send a SupplyCostInquiry to its child stock planners. The children will answer with a SupplyCostReply specifying the cost of delivering parts of the work to be forwarded. Based on the answers, the ForwardWork behaviour assigns work to its children using an AssignWork message. The ForwardWork behaviour will continue until it has completely forwarded the work assignment to its child stock planners.

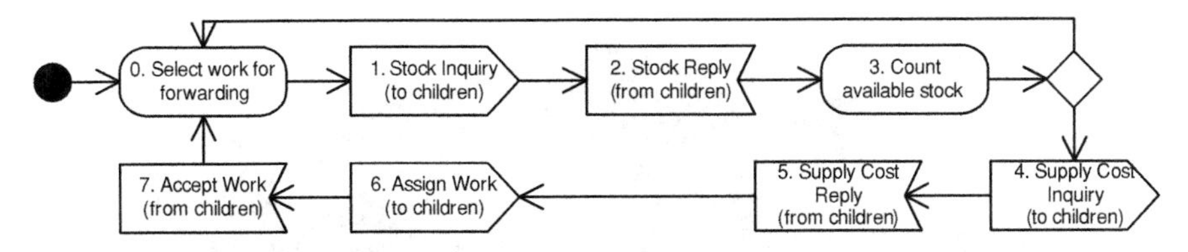

Fig. 2.4 ForwardWork behaviour activity diagram

Note that the ForwardWork behaviour and the AssignWork behaviour trigger each other: the ForwardWork behaviour of a stock planner is triggered by the AssignWork behaviour of its parent and triggers the AssignWork behaviour of its children. The combination of these behaviours allows work to be forwarded from the root of the stock planner hierarchy to the leaf stock planners. From there, it is forwarded to the associated device managers. The latter involves a ForwardWork behaviour that is similar to the one shown in Fig. 2.4.

2.2.2.2 Replenishment Behaviour

The description of the AssignWork behaviour already showed the existence of replenishment work triggered by the forwarding of work through a hierarchy of stock planners. These replenishment assignments are handled by creating delivery work that is forwarded using the ForwardWork and AssignWork behaviours described in Sect. 2.2. Order-driven replenishment may take a long time to be fulfilled, because the required goods need to be transported from one part of the warehouse to another. This may not be acceptable for popular products; these fast movers have to be delivered so often, that they should be in stock all the time. This is achieved by the MinMaxReplenishment behaviour. This behaviour gets triggered when goods have been delivered by a stock planner. The behaviour will check whether the stock level for a certain product has dropped below a minimum. If so, the behaviour will issue replenishment work, which will be handled in the same manner as the order-driven replenishment issued by the AssignWork behaviour. This will ensure that the stock level will be replenished to a specified maximum level.

2.2.2.3 Execution Behaviour

The delivery and replenishment behaviours involve standardised behaviours of the stock planners. The behaviours of the device managers have been standardised as well. An example of a device manager behaviour is the Execution behaviour, which is triggered by an event from a device controller signalling the completion of a task. This behaviour will select the next task to be executed. This involves sending a TaskInquiry message to the consumer device managers; this list contains all tasks that can be executed. The consumers will select a task from this list and send it as

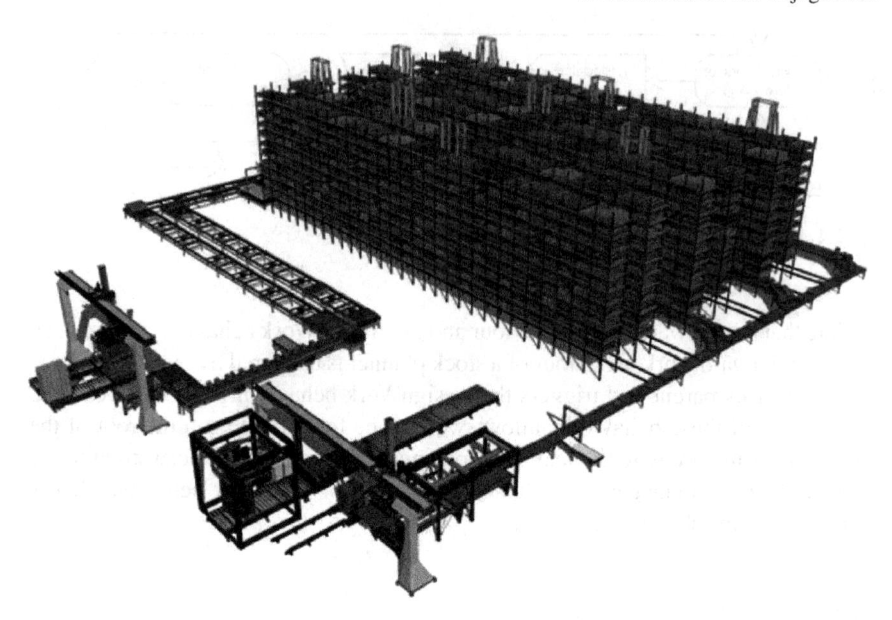

Fig. 2.5 Automated case picking module

a reply. The producer device manager will select one task from the collection of replies and assign it to the signalling device controller, which is responsible for the execution by the material handling system.

2.3 Case Study

We will use the system shown in Fig. 2.5 to illustrate the main ingredients of our reference architecture. It shows an automated case picking module with a palletiser and three case picking cells. Each cell consists of two pick fronts, a reserve, two case pickers, and a tray miniload.

The corresponding component structure is shown in Fig. 2.6. The module, pick fronts, and reserves are stock planners that are part of planning layer; the palletiser, case pickers and tray miniloads are device managers, which reside in the scheduling layer. Besides these, there is an initiation component that plays the role of the ERP system. We do not consider the MFC and MHS layers explicitly.

The module's responsibility is delivering ordered pallets. If it is asked to deliver a pallet, it asks its children, the pick fronts and the reserves, for the corresponding cases. The pick fronts' local stock consists of trays, from which the associated case picker can pick cases. The reserves also store trays of cases, but they cannot be used to deliver cases. If a pick front cannot deliver a certain product, it can be replenished from a reserve: the corresponding tray miniload can pick entire trays and place them in the pick front's local stock.

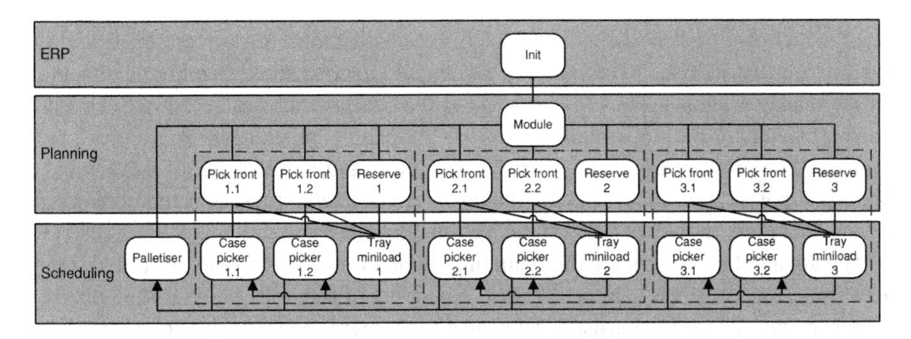

Fig. 2.6 Automated case picking WMCS components

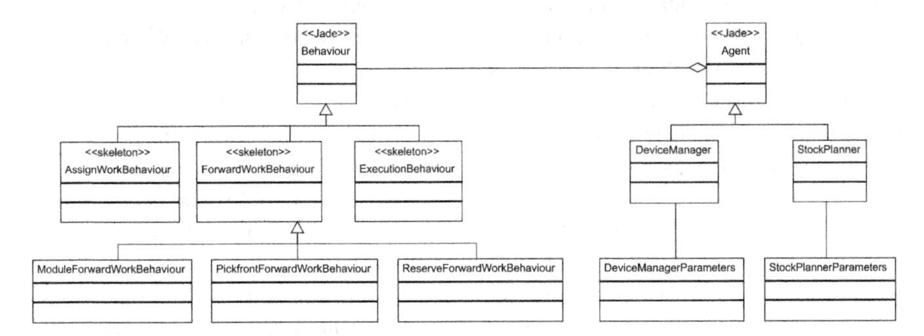

Fig. 2.7 WMCS reference architecture class diagram

Figure 2.6 shows that there need not be a one-to-one relationship between stock planners and device managers. The tray miniloads are associated with two pick fronts and one reserve. It also shows that the device managers are connected in a network, which describes their producer-consumer relationships. Since a tray miniload can replenish two case pickers, it is connected to two case pickers. Similarly, the case pickers are all connected to the palletiser, as they deliver cases for pallet building by the palletiser.

2.4 Implementation

To assess the validity of the WMCS reference architecture introduced in Sect. 2.2, we have implemented a prototype in Java. This prototype implementation only covers the planning layer and the scheduling layer, since these layers cover the most important component behaviours. The actual execution by the MFC and MHS layers is replaced by a simulation, which is included in the scheduling layer. A partial class diagram for the implementation is shown in Fig. 2.7.

The implemented prototype is an agent-based system built upon Jade middleware [1]. A Jade application consists of a number of agents, each having a collection of behaviours. Figure 2.7 shows Jade's Agent class and its Behaviour class, which both have warehouse-specific specialisations in our reference architecture. We distinguish two types of parametrised agents: stock planner agents and device manager agents. These correspond to the components described in Sect. 2.1. The parameters of the stock planners specify the parent and child stock planners, the connected device managers, the initial stock level, and a collection of behaviours. The main parameters of the device managers are its behaviours and the consumer device managers in the network of device managers.

The agents' behaviours are the main ingredients for obtaining the desired system behaviour. As described in Sect. 2.2, the behaviours of the stock planners and device managers have been standardised. This means the interaction between the agents has been fixed, both with respect to the interacting agents and the interface object types. Examples of these generic interaction protocols are described in Sect. 2.2. However, the detailed content of the exchanged messages has not been specified. We will refer to these interaction protocols as *skeleton behaviours*.

Figure 2.8 shows an excerpt of the implementation to illustrate the concept of skeleton behaviours. It shows parts of the code of three classes: Main, ForwardWorkBehaviour, and ModuleForwardWorkBehaviour. The class Main defines the WMCS by specifying the agents and their behaviours. Figure 2.4 shows a behaviour ModuleForwardWorkBehaviour being added to the module agent. This behaviour is a specialisation of the abstract class ForwardWorkBehaviour, which is a skeleton behaviour implementing the activity diagram shown in Fig. 2.4 as a state machine. State 0 of the activity diagram corresponds to state 0 in the code in Fig. 2.8. This state calls an abstract method selectNextWorkAssignment, which corresponds to an empty placeholder function for a business rule in the abstract ForwardWorkBehaviour class. This method is defined in the concrete ModuleForwardWorkBehaviour class.

The notion of skeleton behaviour is further illustrated in Fig. 2.7. It shows the abstract ForwardWorkBehaviour class and three of its application-specific specialisations for the module, the pick fronts, and the reserves. These three behaviours share the interaction protocol defined by the skeleton behaviour ForwardWorkBehaviour, but implement the corresponding business rules differently.

2.5 Experimental Validation

The architecture described in the previous sections is being validated using a series of experiments of the implemented prototype described in Sect. 2.4. The first experiments of the architecture involve the automated case picking module described in Sect. 2.3. The first experiment has been described by Verriet et al. [10]; this experiment focused on stock planner delivery behaviour and device manager execution behaviour. It did not involve replenishment, because the pick fronts were assumed to have sufficient stock to fulfil all orders.

The second experiment focuses on delivery and replenishment behaviour. The experiment starts with pick fronts without any stock and reserves having sufficient stock for all orders. The components of the corresponding WMCS and their behaviours are listed in Table 2.1. There are ten stock planner agents: one module, six pick

```
public class Main {
    ...
    public static void main(String[] args) {
        ...
        moduleAgent.addBehaviour((Behaviour) Class.forName(
            ModuleForwardWorkBehaviour.class.getName()).newInstance());
        ...
    }
    ...
}

public abstract class ForwardWorkBehaviour extends Behaviour {
    ...
    public final void action() {
        ...
        switch (state) {
        case 0:
            retrieveWork();
            break;
        case 1:
            ...
        }
    }
    ...
    private void retrieveWork() {
        ...
        activeWorkAssignment = selectNextWorkAssignment(deliveryWip);
        ...
    }
    ...
    protected abstract WorkAssignment selectNextWorkAssignment(
        Collection <WorkAssignment> workAssignments);
    ...
}

public final class ModuleForwardWorkBehaviour extends
    ForwardWorkBehaviour {
    ...
    protected WorkAssignment selectNextWorkAssignment(
        Collection <WorkAssignment> workAssignments) {
        ...
        return workAssignment;
    }
    ...
}
```

Fig. 2.8 WMCS source code excerpts

fronts, and three reserves. Associated with these stock planners are ten device manager agents: six case pickers, three tray miniloads, and one palletiser. These agents and their connections correspond to those in Fig. 2.6. It shows that each pick front has two device managers: a case picker and a tray miniload. The former allows the delivery of cases; the latter is needed for replenishment. Besides the stock planner and device managers, there is an initiator agent playing the role of the ERP system.

The implementation of the experiment involves a total of 27 skeleton behaviours: 21 stock planner and 6 device manager behaviours. These behaviours have been made concrete for the different agents in Fig. 2.6. This involves a total of 38 ACP-specific behaviours (see Table 2.1). Together these behaviours have 83 business rules overwriting the placeholder functions in the skeleton behaviours. Each behaviour

Table 2.1 Prototype behaviours per component

Component type	# behaviours	# components	# threads
Initiator	3	1	3
Module	5	1	5
Pick front	12	6	72
Reserve	6	3	18
Case picker	6	6	36
Tray miniload	4	3	12
Palletiser	2	1	2
Total	38	21	150

has been deployed as a separate thread using Jade's threaded behaviour factory; this makes a total of 150 behaviour threads running on a single pc. Despite the large number of threads, there is no performance issue, because the threads are all simple programs and only a few threads are active simultaneously.

Figure 2.9 shows a partial trace of the agent conversations in Jade's Sniffer [1]: it shows the messages that are sent between one module, two pick fronts, one reserve, two case pickers, and one tray miniload. We have limited the number of agents to be able to visualise all messages. Figure 2.9 shows three conversations. The first conversation (messages 2.x) involves the module's AssignWork behaviour (see Fig. 2.3). It starts when the module receives an AssignWork message from the initiation agent, which plays the role of the ERP system. As one can see in Fig. 2.9 the AssignWork behaviour causes several messages of different types to be sent, the last one being an AcceptWork reply to the initiation agent. After the first pallet has been assigned to the module, two conversations start running in parallel. The initiation agent forwards the second pallet to the module (messages 3.x) and the module starts to forward this pallet to the pick front and the reserve (messages 4.x).

Figure 2.10 shows a Gantt chart for the execution of the tasks corresponding to two identical pallets consisting of 40 cases each of a unique product. Each colour represents a product. It clearly shows that the pick fronts do not have any stock: before a case picker can pick a case for a pallet, the tray miniload has to replenish the pick front. The case for the second pallet can be picked directly after the first one, because the trays delivered by the tray miniloads contain more than one case. This means that replenishment is needed only once. Figure 2.10 also shows that the work is divided equally over the different case pickers and tray miniloads.

2.6 Architectural Validation

The previous section showed the functional validation of our reference architecture: we were able to forward work through a hierarchy or stock planners and device managers and perform the necessary replenishment. This section considers the validation of the reference architecture in terms of usability aspects. One of the limitations of

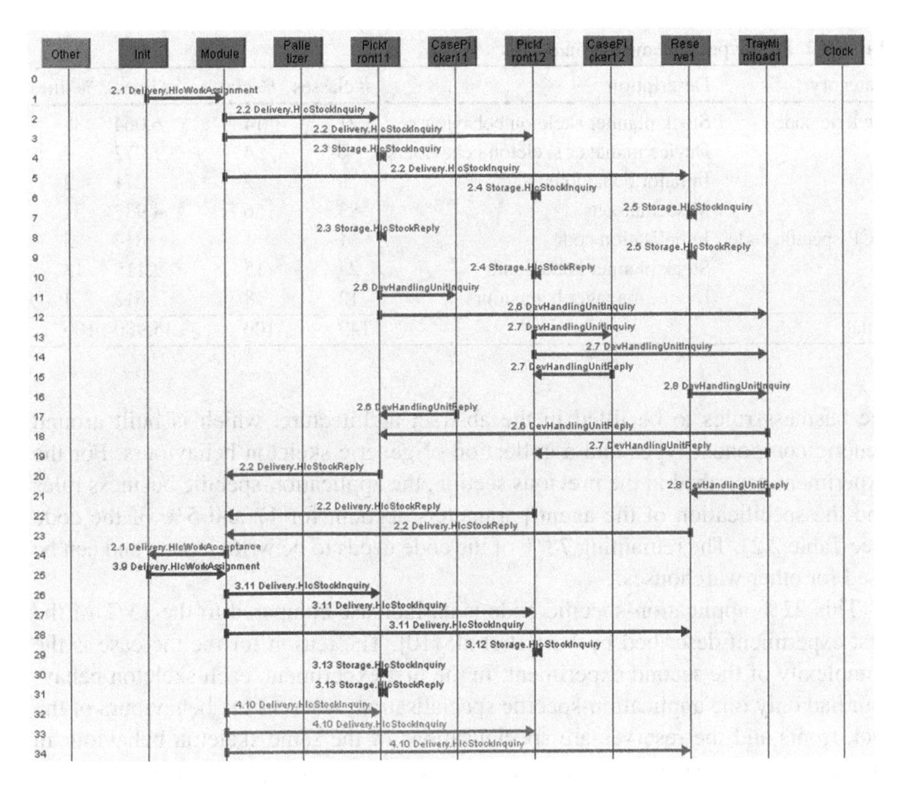

Fig. 2.9 Agent conversations in Jade's Sniffer

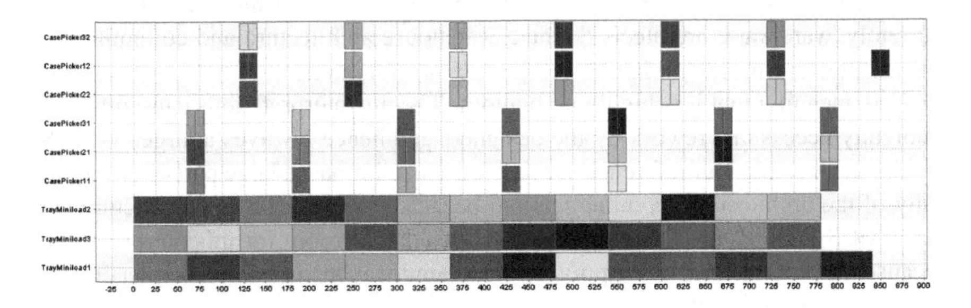

Fig. 2.10 Execution Gantt chart

traditional WMCSs was the limited reusability. This has clearly been addressed by our reference architecture, which is built around generic component types and a collection of generic skeleton behaviours. These generic components can be used and configured for any WMCS.

Reusing the generic components and behaviours has a large positive effect on the WMCS development effort. This effort is limited to the configuration of the system. This involves specifying the connections between the components and defining

Table 2.2 Prototype implementation details

Category	Description	# classes	% classes	# lines	% lines
Generic code	Stock planner skeleton behaviours	21	14	6,064	38
	Device manager skeleton behaviours	6	4	1,072	7
	Initiator behaviours	3	2	271	2
	Miscellaneous	83	56	4,932	31
ACP-specific code	Initialisation code	1	1	814	5
	Stock planner behaviours	23	15	2,115	13
	Device manager behaviours	12	8	612	4
Total		149	100	15,980	100

the business rules to be filled in the abstract architecture, which is built around generic component types and a collection of generic skeleton behaviours. For the experiment described in the previous section, the application-specific business rules and the specification of the agent parameters account for 17 and 5 % of the code (see Table 2.2). The remaining 78 % of the code needs to be written once and can be used for other warehouses.

This 22 % application-specific code is an increase compared to the 15 % of the first experiment described by Verriet et al. [10]. The reason for the increase is the complexity of the second experiment. In the first experiment, each skeleton behaviour had only one application-specific specialisation, whereas the behaviours of the pick fronts and the reserves are specialisations of the same skeleton behaviour in the second experiment. This suggests that the amount of application-specific code strongly depends on the number of WMCS components.

There is more to the WMCS development effort than just the amount of code. Currently, warehouse architects define a warehouse architecture and communicate this architecture to the WMCS developers. This communication has not been formalised, meaning that it is highly ambiguous. The misinterpretations caused by the ambiguity necessitate rework by the designers and hence involves a larger WMCS development effort. Our reference architecture can eliminate a large part of the ambiguity of the architectural communication, because it provides a clear structure for the WMCS architecture. This structure will leave less room for misinterpretation, because the architectural description will contain local business rules with a clearly defined scope and interface.

Because of the standardised interfaces, the components of our architecture are also exchangeable. A change in the stock planner hierarchy can easily be accommodated by altering the parent-child relationships in the WMCS configuration. The new structure may require an alteration of business rules, but this is limited to the components affected by the structural change. Similarly, a change of the network of device managers does not involve a large redesign effort.

To improve the WMCS development even further, we are currently developing a graphical editor (see Fig. 2.11), in which the warehouse architect can select the required warehouse components and their interfaces [6]. Moreover, he will be able to select the components' behaviours and specify the appropriate application-specific

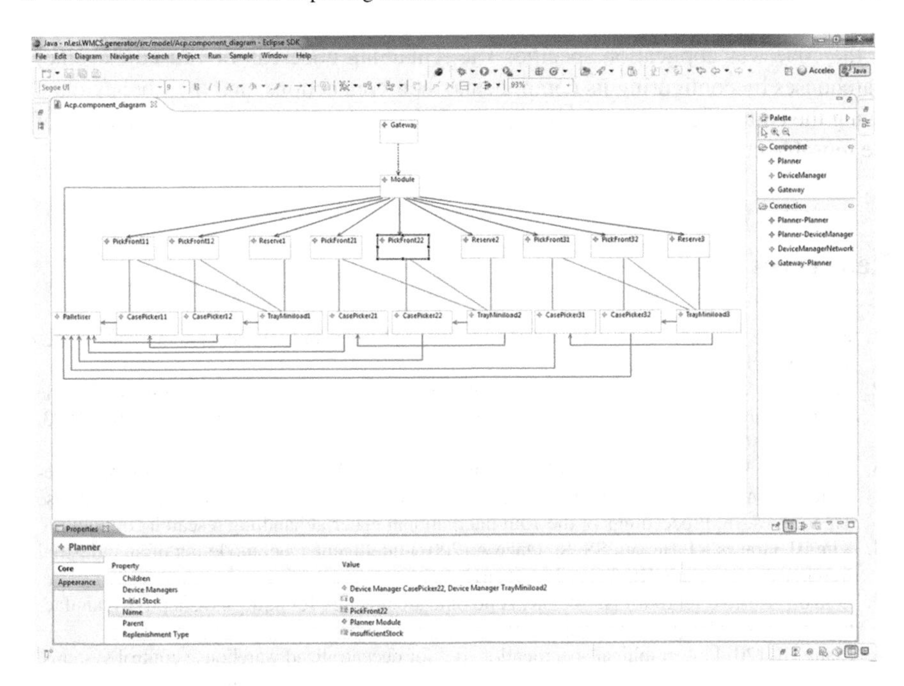

Fig. 2.11 Prototype WMCS editor

business rules. A large part of a WMCS code can be generated automatically from the architect's specification using a model-driven software engineering approach (see Chap. 4). The generated code includes the code needed to define the components and their parameters. The remaining code, the application-specific business rules, is to be written by the WMCS designers.

2.7 Conclusion and Outlook

In this chapter, we have described a WMCS reference architecture. This architecture defines the generic WMCS components, their connections, and their generic behaviours. A WMCS can be designed by specifying the parameters of the components and behaviours. The behaviours' parameters are application-specific business rules, which create concrete behaviours from abstract skeleton behaviours. First experiments successfully demonstrated forwarding of work through a hierarchy of components and order-driven replenishment of local stock.

The experiments have also shown that the reference architecture can greatly reduce the WMCS development effort. The architecture provides a communication framework, which reduces the chance of ambiguity by limiting the scope to local business rules. Moreover, a lot of generic code can be reused: in our experiment, only 22 %

of the code was application specific. The remaining code can be reused for other warehouses by configuring its components, their connections, and their behaviours. Using the editor we are developing, part of the application-specific code and all of the reusable code can be generated automatically.

References

1. Bellifemine F, Caire G, Greenwood D (2007) Developing multi-agent systems with JADE. Wiley, Chichester
2. Fleetwood M, Kotak DB, Wu S, Tamoto H (2003) Holonic system architecture for scalable infrastructures. In: IEEE international conference on systems, man and cybernetics 2003, vol 2, pp 1469–1474
3. Graves RJ, Wan VK, van der Velden J, van Wijngaarden B (2008) Control of complex integrated automated systems—system retro-fit with agent-based technologies and industrial case experiences. In: Proceedings of the 10th international material handling research colloquium
4. Kim BI, Graves RJ, Heragu SS, St. Onge A (2002) Intelligent agent modeling of an industrial warehousing problem. IIE Trans 34:601–612
5. Leitao P, Restivo F (2006) ADACOR: A holonic architecture for agile and adaptive manufacturing control. Comput Ind 57:121–130
6. Liang HL (2011) A graphical specification tool for decentralized warehouse control systems. SAI Technical Report, Eindhoven University of Technology, Eindhoven
7. Moneva H, Caarls J, Verriet J (2009) A holonic approach to warehouse control. In: 7th international conference on practical applications of agents and multi-agent systems (PAAMS 2009), Advances in intelligent and soft computing vol 55. Springer, Berlin, pp 1–10
8. Ten Hompel M, Schmidt T (2006) Warehouse management: Automation and organisation of warehouse and order picking systems. Springer, Berlin
9. Van Brussel H, Wyns J, Valckenaers P, Bongaerts L, Peeters P (1998) Reference architecture for holonic manufacturing systems: PROSA. Comput Ind 37:255–274
10. Verriet J, van Wijngaarden B, van Heusden E, Hamberg R (2011) Automating the development of agent-based warehouse control systems. In: Trends in practical applications of agents and multiagent systems, Advances in intelligent and soft computing, vol 90. Springer, Berlin, pp 59–66

Chapter 3
Decentralised Warehouse Control through Agent Organisations

Huib Aldewereld, Frank Dignum, and Marcel Hiel

Abstract Warehouse management and control systems are traditionally highly optimised to a specific situation and do not provide the flexibility required in contemporary business environments. Agents are advocated to provide adaptiveness and flexibility, and have been used to solve specific problems in the warehouse logistics domain. While decentralising the warehouse control by using agents solves issues with modularity and reusability of the software, the complexity of the system may increase, decreasing the predictability of the system. In order to solve these issues and bring the agent-based control on par with centralised control, we propose the use of agent organisations.

3.1 Motivation

Warehouse management and control systems (WMCS) are traditionally centralised, monolithic software systems that are highly optimised for a specific situation. However, in the current business environment, where mergers, acquisitions and rapid product development happen frequently, companies are in a continuous state of flux. The warehouses used by these companies (as customer or owner) are therefore subject to a lot of changes. Examples of such changes range from withdrawal or

H. Aldewereld (✉) · F. Dignum · M. Hiel
Department of Information and Computing Sciences, Utrecht University,
P.O. Box 80089, 3508 TB Utrecht, The Netherlands
e-mail: huib@cs.uu.nl

F. Dignum
e-mail: dignum@cs.uu.nl

M. Hiel
e-mail: hiel@cs.uu.nl

R. Hamberg and J. Verriet (eds.), *Automation in Warehouse Development*,
DOI: 10.1007/978-0-85729-968-0_3, © Springer-Verlag London Limited 2012

addition of (types of) products, slow-moving products becoming fast moving (and vice versa) to the addition or removal of hardware.

The hardware that is used in warehouses has been subject to evolution which resulted in an increasingly modularised approach to decomposing the warehouse tasks and improving reusability. This evolution is, however, not reflected in the software that controls these machines and many WMCSs are still centralised and monolithic. While performance will remain an important aspect in WMCS development, progress has to be made in increasing modularity, reusability, and exchangeability if WMCS are to evolve alongside the component-based hardware developments.

Recently, multi-agent systems (MAS) were proposed to introduce more flexibility in warehouse control. Agents, being software "entities" characterised by properties such as autonomy and pro-activeness, serve as an alternative to the centralised approach, potentially alleviating problems in modularity, reusability, and exchangeability. Typically, every hardware component is linked to a specific (type of) agent that is made responsible for the correct functioning of that component, whereas the collection of these agents is responsible for the correct functioning of the warehouse as a whole.

The decentralised nature of autonomous agents improves the modularity of a WMCS over a centralised approach. Moreover, the decentralisation, combined with standardised interfaces between the agents, can lead to an improvement of reusability and exchangeability. However, it also decreases the system's performance and predictability; the agent system is harder to optimise to specific situations, and the autonomy of the agents, which leads to better flexibility, also make it harder to predict system behaviour. The main problem here is that aspects such as efficiency, flexibility, and robustness are aspects that pertain to the system as a whole. One cannot optimise all three at the same time, but one has to balance the three aspects. Due to the distributed nature of the agent system, it is harder to guarantee this overall balance.

In order to prevent anarchy and regulate the agents within a control system, we propose the use of agent organisations. Agent organisations are a control mechanism to direct the agents in a system to behave in a preferred manner. Agent organisations are also used to specify exactly those aspects of the system that need to be guaranteed by the agents together. Before we discuss the functioning of the organisations, we first describe the architecture of the control system.

As example and test domain we use retail warehouses as introduced in Chap. 1. We focus on the storage and order handling components of the warehouse. Products are packed and/or placed in boxes or containers, generally referred to as Transport and Storage Units (TSU). The types of components in our warehouse are miniloads, conveyor belts, and (order-picking) workstations. The miniloads are storage units where TSUs are kept, the conveyor belts are responsible for transporting TSUs from miniloads to workstations and back, and the workstations are the places where operators pick products from TSUs for fulfilling orders.

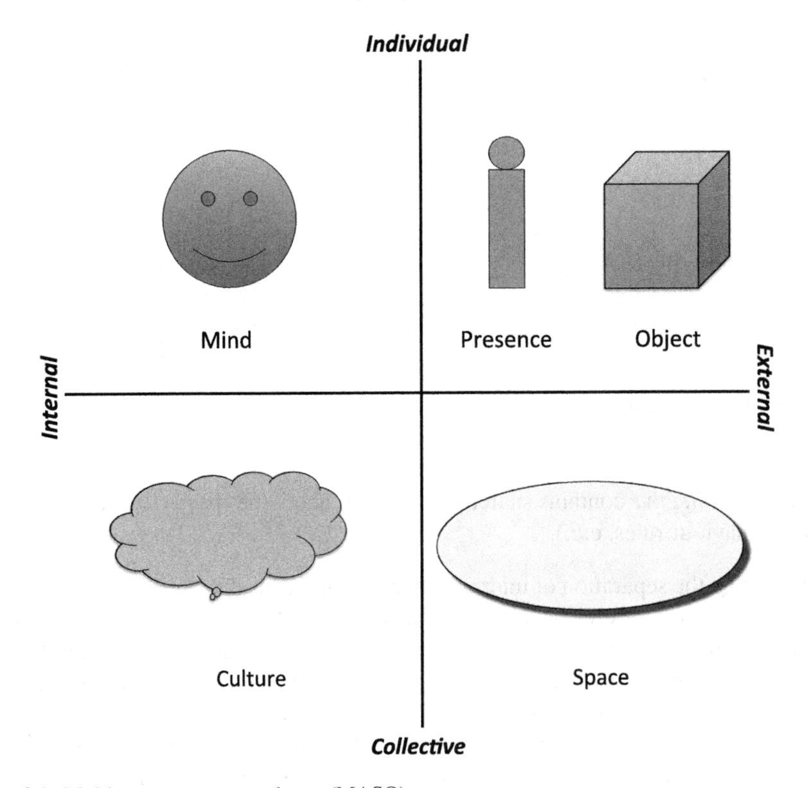

Fig. 3.1 Multi-agent system quadrants (MASQ)

3.2 Modelling Agent-Based WMCS

The architecture is based on the MASQ meta-model [6] for structuring basic elements that compose the interaction processes of agents, environments, and organisations. MASQ is an acronym of Multi-Agent System Quadrants and defines four perspectives on agent-based interaction, according to two orthogonal axes, introducing a separation of internal/external perspectives on the one side and individual/collective perspectives on the other side. The axes and perspectives of MASQ are shown above in Fig. 3.1. The four quadrants of MASQ are:

- *Internal-Individual* containing the minds of the agents (the mind contains the beliefs and desires of the agent and has the reasoning capabilities);
- *External-Individual* containing presences and objects (presences are objects that can be embodied by a mind; presences can be physical, like an order-picking workstation, or virtual, like a planning presence);
- *External-Collective* contains spaces (a space is an environment that defines the interactions between the contained presences; presences can be in only one space, but a mind can control more than one presence (in various spaces));

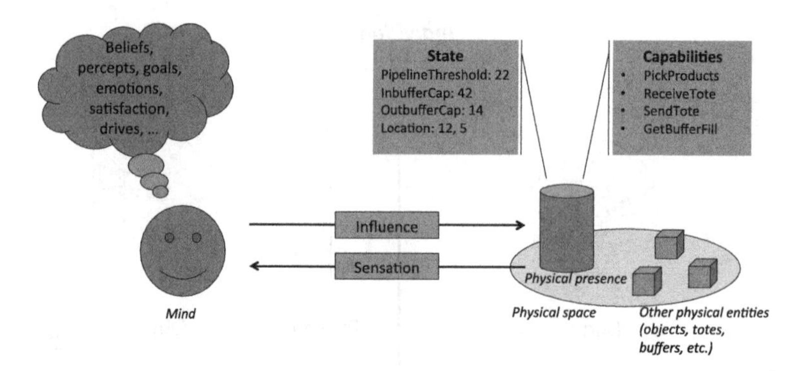

Fig. 3.2 MASQ mind-presence relation

- *Internal-Collective* contains shared conceptions about the spaces (like norms, culture, behaviour rules, etc.).

In MASQ, the separation of internal and external means that an agent has a single *mind* but can have several *presences*. A presence can be either a physical, virtual, or social representation of the agent in an environment. The environment, combined with the rules and protocols that the agents use to interact, is called a *space*. The presences are the means of an agent to interact with its environment (space), and presences of other agents in that same space. The mind corresponds to the internal structure of an agents and also serves as the "glue" between the different presences. The interaction between the different parts of the MASQ meta-model is shown in Fig. 3.2. The figure depicts the interaction between a *Picking* agent's mind and its physical presence. The mind contains the knowledge about the objectives of the agent and what actions to perform (based on sensations (inputs) from this and other presences it embodies). The presence is the physical representation of the agent in the world (in this case, an order-picking workstation component), which has a state (describing, for instance, its location in the physical space, how many totes it can contain in its buffers, etc.) and specific capabilities (the actions it can perform).

3.3 The Architecture

Warehouse management and control systems are typically thought of in three layers of operation, namely the plant (execution), scheduling, and planning. This creates a clear separation of responsibilities, which makes it easier to create a modular design and thereby support decoupling of the different aspects. This separation is reflected in the proposed architecture (see Fig. 3.3). Each of the layers of the warehouse management and control system (plant, scheduling, and planning) corresponds to a space in the external-collective perspective of the MASQ model. Each space defines the constraints on the interaction (e.g. the organisation or hardware configuration),

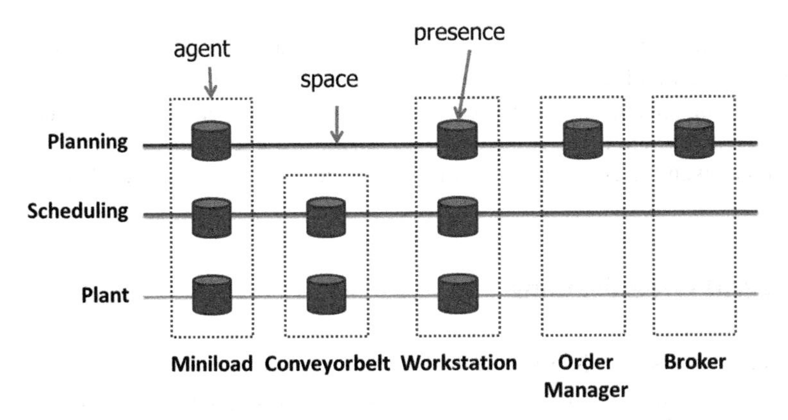

Fig. 3.3 Presences of the agents in a warehouse

interaction protocols, and incorporates its own presences with implemented business rules, thereby improving modularity and maintainability.

The presences of each of the agents is represented by a box, an agent is represented by a dotted line around a group of boxes. For instance, the miniload agent has three presences, one on each layer, indicating that the mind of the miniload controls a presence in each of the spaces. Some agents do not have presences on all the layers, indicating that they cannot participate in interactions on that layer. For instance, the conveyor belt agent is not involved in the planning, and only has scheduling and plant presences. Planning is all about deciding on order fulfilment (which order-picking workstation is going to pick a particular order) and replenishing components for that purpose (which miniload is going to provide the product totes needed for picking that order), and since the travel time from the miniloads to the order-picking workstations is static in this decision process, the participation of the conveyor belt (which is responsible for transporting the totes from the miniloads to the workstations, and back) is not required in the planning interaction.

Business rules (i.e. the decision rules of the agent) are located and handled within the respective presences of the agent. Business rules are closely related to the functions that implement the decision. Therefore placing the business rules in the presences creates a modularity per implementation of a business rule. For example, two types of conveyor belt schedule presences can be implemented, one that supports only FIFO and one that supports more intelligent scheduling. Thus, changing the behaviours of the agents then becomes as simple as switching between different presences of that agent without the need to re-implement the decision elements of the agent itself.

Having the business rules in the presences and not in the mind has two additional advantages. First, it supports simultaneous development in different spaces as the overlap (in the mind) is minimised. Second, it prevents the agent's mind from becoming more and more complex with each presence that it controls. This complexity would impair any potential flexibility and adaptability of the agents. Instead, using the minds only as the "glue" that keeps all the presences of the agent together means that the minds can be designed generically and reused for each type of agent (since the type-specific decision rules are within its presences). The minds just trigger the tasks that need to be performed in the separate presences and any decisions that have to be taken for executing these tasks are handled in the presences themselves.

3.4 Agent Organisations

An agent organisation is a structure that supports the overall model of a multi-agent system. This model restricts the behaviour of the agents functioning in that organisation and ensures that overall objectives are met. In the architecture described in the previous section, the interactions on the planning and scheduling spaces are governed by an agent organisation. As the plant is solely concerned with controlling the hardware, we do not have an organisational model for the plant space, but use (physical) restrictions from the hardware layout as constraints on the (plant) interactions instead. This is depicted in Fig. 3.4. The small coloured boxes on the conveyor belt represent moving TSUs, their colour indicates the product family to which the products in the TSU belong. The squares with arrows on them represent buffers, where the arrow indicates the direction of movement.

The organisational models used in our architecture are based on the OperA framework [3, 4], which makes a separation between organisational roles and agents. Roles in OperA are like positions described on a high-level of abstraction, focusing on what is required from that position instead of the specific capabilities of agents that may fill that position. This makes the organisational specification applicable in a wider context (it does not matter what type of order-picking workstations are used) and easier to maintain. OperA contains the following aspects:

Social: The social aspect involves roles, objectives, and dependencies between roles based on objectives. Roles identify the important parties required for achieving organisational objectives and enable the developer to abstract away from the specific actors (the presences in the concrete sense, or the agents in an abstract sense) that will perform the required activities. Roles are dependent on each other for the achievement of their objectives and therefore require interaction.

Interaction: In systems containing multiple components, interaction is an important aspect. Through interaction the objectives of the organisation are realised. The interaction in OperA is structured in scenes where each scene represents the parties that are involved for reaching a certain objective (e.g. order assignment, replenishment, etc.). The interaction is structured using landmarks. A landmark describes the state that should be reached during the interaction. By specifying only landmarks, the actual protocols may vary (as long as they achieve the same results).

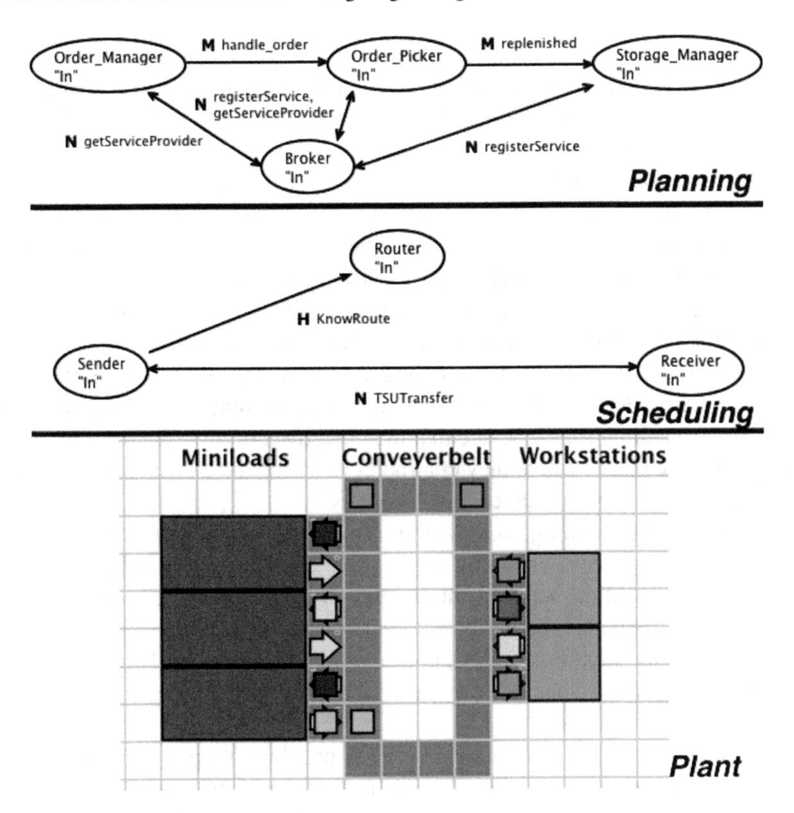

Fig. 3.4 Models per layer. *Top:* Planning organisation model. *Middle:* Scheduling organisation model. *Bottom:* Hardware configuration

Normative: Norms specify how actors ought to behave in certain situations. Business rules are one way of implementing norms. One of the most distinctive features of norms is that agents might violate them if circumstances require. For instance, if orders should be handled in a FIFO ordering, an agent might deviate from this rule in case one of the products for that order is not in stock or when a priority order arrives. Within the framework one also describes how the system should handle the violation situations (which lends the system its flexibility).

To get some further clarity on how the organisation functions, let us focus on the planning layer organisation. There are several design decisions in the agent-based planning approach for warehouse management and control. These can be based on three principal questions: (1) Who talks to whom? (2) Who to talk to for delivery/replenishment? and (3) How to talk?

Who talks to whom? This question is answered directly via the organisational model. The organisation defines the important parties (roles), and the way these roles should interact in order to reach the organisation's objective. The top part of Fig. 3.4 illustrates the roles in our warehouse planning organisation. These roles, partially, coincide with the types of hardware components distinguished in the hardware layout (bottom part of Fig. 3.4); the idea being that the roles represent some abstract parent class of the hardware components that have the same function (i.e. they all achieve the same objective). For example, a manual order-picking workstation or an automated order-picking workstation both achieve the objective of order picking, and can thus both be abstracted into a single role, namely the Order_Picker. The objectives of the roles in the planning space indicate what type of things will be optimised by the different roles. The Order_Picker role might have the objective to pick as many TSUs as possible. The Storage_Manager role might have the objective to minimise storage use or alternatively to maximise the throughput of TSUs. The Order_Manager is responsible for handling orders and thus will try to minimise the handling time of orders (or the average handling time or maybe the deviation of the average order handling time). Notice that the objectives of different roles might not all be in line. The places where they contradict is where choices have to be made in their interaction on how to balance the objectives.

The arrows in the planning organisation (top part of Fig. 3.4) indicate the dependencies between the roles, where the origin of an arrow typically indicates the initiator of the conversation between those roles. By defining these roles, we allow an agent, and thereby also the hardware component it controls, to adapt a position appropriately for the situation. For example, a miniload may not only play the role of Storage_Manager but when necessary also that of Order_Picker when full TSUs are required to fulfil orders.

Who to talk to for delivery/replenishment? A problem of open dynamic systems is how agents know who they should talk to for a particular service. As agents may join and leave the system (e.g. component breakdown, adding new hardware) the provider for a particular service (deliver, replenish, relocate) can change. This makes it necessary to maintain and keep up-to-date information on who provides what service(s). This information is stored and provided by the broker. Although hardware components of a warehouse are typically not subject to frequent changes, by providing the controlling software the capability of handling different situations, performance during abnormal situations can be improved dramatically. For example, by incorporating the broker, our system allows hardware components to be taken off-line (for instance, for maintenance) and keep a (albeit reduced) steady performance.

How to talk? Besides determining which agent/role to talk to, it is important to decide how the agents talk to each other. This aspect is also covered by the organisational model through the use of the dependencies. OperA distinguishes three types of dependencies, each resulting in a different type of interaction: **H**ierarchy dependencies (indicated by the capital **H** next to the dependency in Fig. 3.4) specify that the interaction is of a delegation type. The initiating role delegates the responsibility of achieving the objective to the sub-ordinate role. **N**etwork dependencies (indicated with the capital **N**) specify that the roles need to coordinate to achieve the objective (typically implemented as a request-inform type of communication). The interactions between the planning roles and the broker are typically of this type. **M**arket dependencies (indicated by a capital **M**) specify that the roles have to use bidding (or auction) kind of protocols to achieve the objective of the dependency.

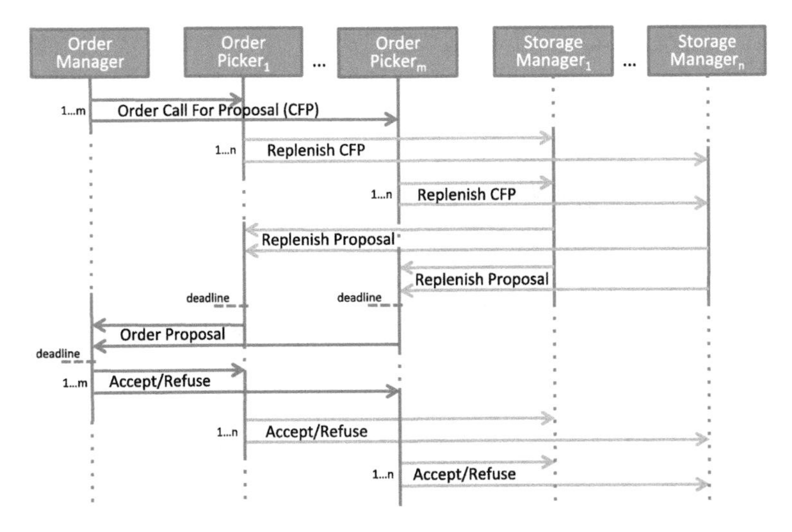

Fig. 3.5 Order planning communication

Examples of the Market dependencies in the planning organisation are the interactions between the Order_Manager and the Order_Picker with respect to order delivery, and the interactions between the Order_Picker and Storage_Manager with respect to replenishments. Market dependencies are implemented in our architecture through the Contract Net protocol (CNP; a standardised bidding protocol, see [5]). An overview of the implementation of those two market dependencies in CNP is show in Fig. 3.5.

To find the best component to fulfil a newly arrived order, a *call for proposal* is sent to all providers of the delivery service (which are the Order_Picker s in our example). The pickers answer with a *proposal* containing the amount of time it would take that picker to complete the order. The Order_Manager weighs the received proposals and sends an *accept-proposal* to the picker that offers the best proposal. The other pickers receive a *reject-proposal* message. If pickers require resupplies to fulfil the order, it can start a replenish interaction with the Storage_Manager s (which is also handled as a CNP). This interaction can interleave the delivery interaction, as shown in Fig. 3.5 (delivery interaction is coloured red, replenish interaction is coloured green).

The use of market relations for the service requests means that, given appropriate bidding and weighing mechanisms, balance of the workload of the components is achieved automatically.

3.5 The Simulator

A simulator has been developed in ABC Lab (the implementation of MASQ within Repast Simphony [7]) to show the strengths (and investigate the weaknesses) of the agent-organisation approach. The simulator allows loading and testing different

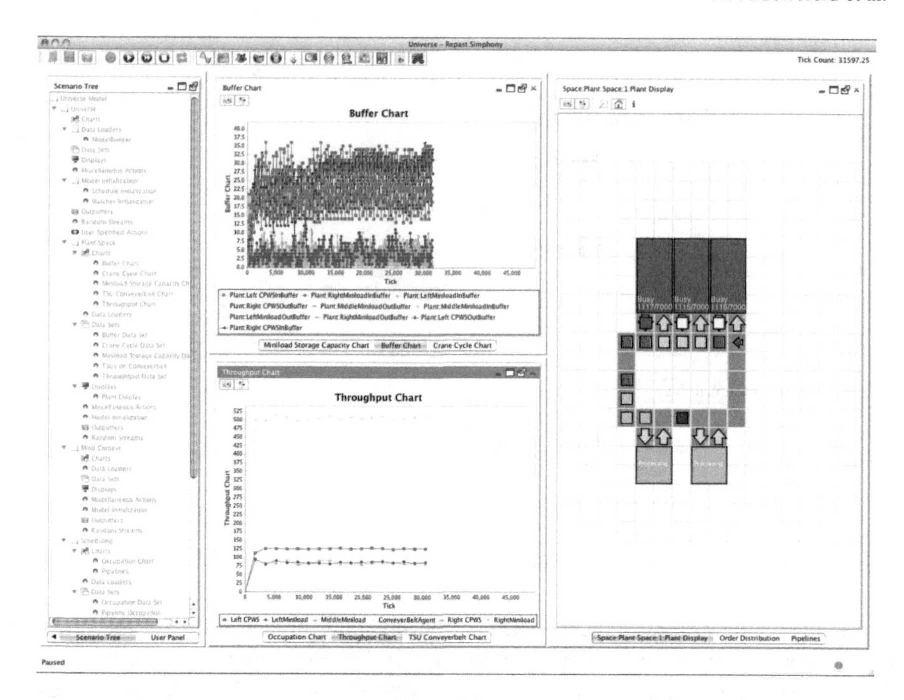

Fig. 3.6 ABC lab warehouse management simulator

scenarios where the presences of agents can be interchanged. Since the business rules of the agents are incorporated in the presences of the agents, this allows testing combinations of different components and business rules in a quick manner (without the need to re-write the control structure all over again). The ultimate goal for this simulation tool is to be usable as a design aid to allow making changes to the hardware and organisation settings and be able to test the results on the performance of the system overall. A screenshot of the simulator in progress (with three miniloads and two order-picking workstations) is shown in Fig. 3.6.

The simulator automatically tracks various aspects of the system to check the performance, robustness, and flexibility. For example, the simulator creates real-time graphs of the system's throughput (TSUs handles per hour), the pipeline fill rates (the amount of work assigned to each component per time), and the distribution of the orders over the components responsible for the delivery service. A video of the simulator can be found at [1].

3.6 Concluding Thoughts and Outlook

By adding an organisation on top of the MAS, we maintain the advantages gained by introducing agents (decentralisation, modularity, etc.), but lower (or even eliminate) some of the disadvantages. The organisation explicitly represents the required/ideal

Table 3.1 Comparison between approaches for designing WMCS

	Centralised	Agents	Agent organisations
Modularity	None	High	High
Reusability	Low/none	Average/high	High
Performance	High	Average	Average/high
Robustness	Low/none	Average	High
Predictability	High	Low/average	High
Flexibility	Low/none	High	High

behaviour of the agents, thus improving the system predictability. The organisation also better controls the performance of the system, as this is a shared (organisational) goal of all the agents in the organisation. And furthermore, lots of the interaction and coordination specification of the system can be moved to the organisation, making the agents simpler and less situation specific, thus further increasing the reusability of the system. Table 3.1 above summarises the comparison.

The simulations of the organisation-based WMCS have been shown to be performing on par with existing retail warehouse control (simulations) based on the same (real-world) order data used by both simulations. The main results of this research are, however, on improving the modularity of the solution, which is indicated by the ability to quickly change the warehouse composition and agent's behaviours without having to re-write the control system. The advantages of this (over other control systems) is, however, harder to prove.

Organisations try to achieve three global objectives, namely efficiency, robustness, and flexibility. An organisational structure maintains a certain balance between these objectives with respect to its environment. However, if the environment changes, the balance may shift into an unfavourable direction and a re-organisation may be required. An exhaustive overview of changes and corresponding actions for adaptation is hard to provide. The proposed architecture, due to its separation of layers and explicit modelling of interactions per layer, allows the specification of change patterns [2]. Change patterns, and design patterns in general, provide a structured documentation thereby making the transfer of knowledge between developers easier. Making this knowledge explicit thus reduces the time required for (re-)design and (re-)implementation.

We are currently working on an automated way of transforming OperA organisational models into simulations. Because of the extensive use of models in every step of the development (and for every layer of the WMCS), such an (semi-)automated translation from model to simulation (or even the implementation of a WMCS) could be achieved with the help of techniques from Model-Driven Software Engineering (MDSE; see Chap. 4). With a model-to-model transformation from the OperA organisation model to the ABC Lab model (the model of the simulation environment used) and a subsequent model-to-code transformation to generate the simulator code, changes to the organisation and the agent presences can be (currently done manually) implemented and tested easier and quicker.

References

1. Aldewereld H, Dignum F, Hiel M (2011) Falcon agent-based WMCS simulator. http://www.youtube.com/watch?v=R4CCMsCZCbY, Viewed May 2011
2. Aldewereld H, Dignum F, Hiel M (2011) Re-organization in warehouse management systems. In: Proceedings of the IJCAI 2011 workshop on artificial intelligence and logistics (AILog-2011), pp 67–72
3. Aldewereld H, Dignum V (2011) OperettA: Organization-oriented development environment. In: Proceedings of the 3rd international workshop on languages, methodologies and development tools for multi-agent systems
4. Dignum V (2004) A model for organizational interaction: based on Agents, founded in Logic. SIKS Dissertation Series 2004-1. Utrecht University, Utrecht
5. FIPA (2002) FIPA contract net interaction protocol specification. http://www.fipa.org/specs/fipa00029/SC00029H.html, Viewed May 2011
6. Stratulat T, Ferber J, Tranier J (2009) MASQ: towards an integral approach to interaction. In: Proceedings of the 8th international conference on autonomous agents and multiagent systems, 2:813–820
7. Tranier J, Dignum V, Dignum F (2009) A multi-agent simulation framework for the study of intermediation. In: Proceedings of the 6th conference of European social simulation association

Chapter 4
Model-Driven Software Engineering

Marcel van Amstel, Mark van den Brand, Zvezdan Protić, and Tom Verhoeff

Abstract Software plays an important role in designing and operating warehouses. However, traditional software engineering methods for designing warehouse software are not able to cope with the complexity, size, and increase of automation in modern warehouses. This chapter describes Model-Driven Software Engineering (MDSE), a discipline aimed at dealing with the increased complexity of software by focusing on the problem domain rather than on the solution domain. In warehouse design, this is achieved by using formal models to describe warehouses, and by using model transformations to transforms those models to, e.g. source code. MDSE relies on tools more than traditional software engineering. Thus, to reap the full benefits of MDSE, tools for developing, managing, and transforming models should be designed, implemented, used, and validated.

4.1 Introduction

The behaviour of systems that comprise a warehouse becomes increasingly complex and requires a higher level of automation due to more advanced business processes of customers. At the same time, high quality standards should be attained. There-

M. van Amstel (✉) · M. van den Brand · Z. Protić · T. Verhoeff
Department of Mathematics and Computer Science, Eindhoven University of Technology, P.O. Box 513, 5600 MB Eindhoven, The Netherlands
e-mail: M.F.v.Amstel@tue.nl

M. van den Brand
e-mail: M.G.J.v.d.Brand@tue.nl

Z. Protić
e-mail: Z.Protic@tue.nl

T. Verhoeff
e-mail: T.Verhoeff@tue.nl

R. Hamberg and J. Verriet (eds.), *Automation in Warehouse Development*,
DOI: 10.1007/978-0-85729-968-0_4,

fore, software plays a pivotal role in today's and tomorrow's warehousing industry. Surveys have shown, however, that traditional software development methodologies are unable to cope with the size and scope of projects required by large companies [14, 6]. One reason for this is that the transfer of knowledge between different phases of the development process is problematic. Design decisions that have been made in one phase need to be manually interpreted in the next. For example, the detailed layout of a warehouse is given to software engineers who subsequently need to develop the software for controlling the equipment in that warehouse. Since the development teams among which knowledge has to be transferred have different backgrounds, this may lead to all kinds of *misinterpretations*.

There are two reasons for this problem. First, the formalisms used in different stages of the development process may have different semantics and thereby different expressive power. Therefore, it may occur that concepts expressed in one formalism cannot be expressed in the other. In an attempt to bridge these semantic gaps, a slightly different interpretation may have to be chosen for certain concepts. Second, design decisions tend to be insufficiently documented. For example, some decisions that are considered to be trivial for a development team may have been undocumented. In this case, developers in a subsequent step may interpret these trivialities in a different way than intended. Surveys show that these problems may result in a situation where maintenance of the software accounts for up to 90% of the total cost of the software [3].

Model-driven software engineering (MDSE) is an emerging software engineering discipline intended to improve traditional software development methodologies. MDSE aims at managing software complexity and improving productivity [11]. This is achieved by providing means to raise the level of abstraction from the solution domain to the problem domain, and to increase the level of automation in the development process. Raising the level of abstraction is achieved by employing domain-specific languages (DSLs) that offer—through appropriate notations and abstractions—expressive power focused on, and usually restricted to, a particular problem domain [26]. A DSL enables system engineers to model in terms of domain concepts rather than concepts provided by general purpose formalisms, which typically do not provide the required or correct abstractions. For example, in Sect. 4.3 we describe a DSL for material flow diagrams. This DSL offers the concepts to describe a warehouse system in terms of material flows, rather than in terms of general concepts like sets and lists.

The use of (mostly graphical) domain-specific models leads to a better understanding of the system by the various stakeholders. Automating the transition between different design process steps is achieved by using model transformations. Model transformations provide an automatic mechanism to generate new models from existing models, or to update those models. This facilitates the transfer of models between phases of the software development life-cycle, while ensuring consistency between models. By using model transformations, models do not have to be interpreted by humans, which, in combination with the domain-specific abstractions, greatly diminishes the risk of misinterpretations. Moreover, since the process is automated, it is quicker and less error-prone. Therefore, MDSE is a promising approach to deal with the increasing level of automation and complexity that the warehousing industry is

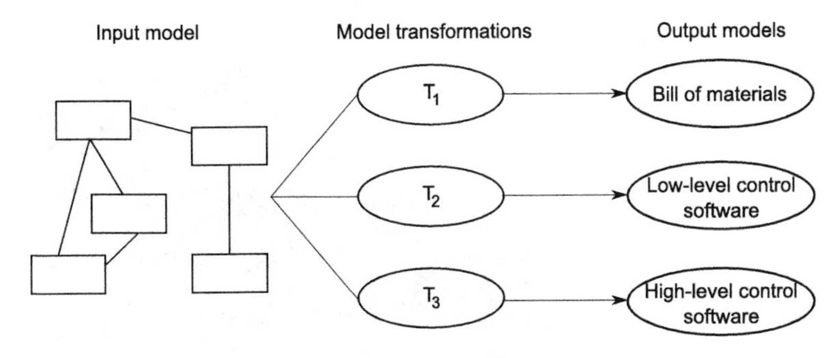

Fig. 4.1 Example of an MDSE process

facing. An example of this process is shown in Fig. 4.1. A model of part of a warehouse is transformed into various implementation artefacts. In this way, consistency between the output models is ensured, since they are all generated from the same input model.

The remainder of this chapter is structured as follows. Section 4.2 presents a case study that shows how MDSE can be employed in a warehousing environment. We provide another case study in Sect. 4.3 and shortly reflect on the case studies in Sect. 4.4. However illustrative, these are small-scale case studies. To enable a model-driven approach in large-scale projects, more enabling technology is required. In Sects. 4.5 and 4.6, we discuss such technology. Section 4.5 concerns the comparison of models and Sect. 4.6 deals with analysing (the quality of) model transformations. Conclusions are given in Sect. 4.7.

4.2 Code Generation by Model Transformations: A Case Study

In this case study, we specified the behaviour of the three-belt conveyor system depicted in Fig. 4.2 using a DSL [19, 20]. The DSL we designed and implemented is called *Simple Language of Communicating Objects* (SLCO). SLCO has an intuitive graphical syntax to describe the structure and behaviour of a system and offers constructs to make models concise. The structure of a system is described using classes, objects, ports, and channels. The behaviour of a system is described using state machines. Figure 4.3 depicts the state machines that control the conveyor system of Fig. 4.2.

4.2.1 Code Generation by Model Transformation

We implemented a chain of model transformations that transforms SLCO models into NQC, a restricted version of the C programming language [1]. The semantic properties of the NQC implementation platform differ from those of our DSL. This

Fig. 4.2 Conveyor system

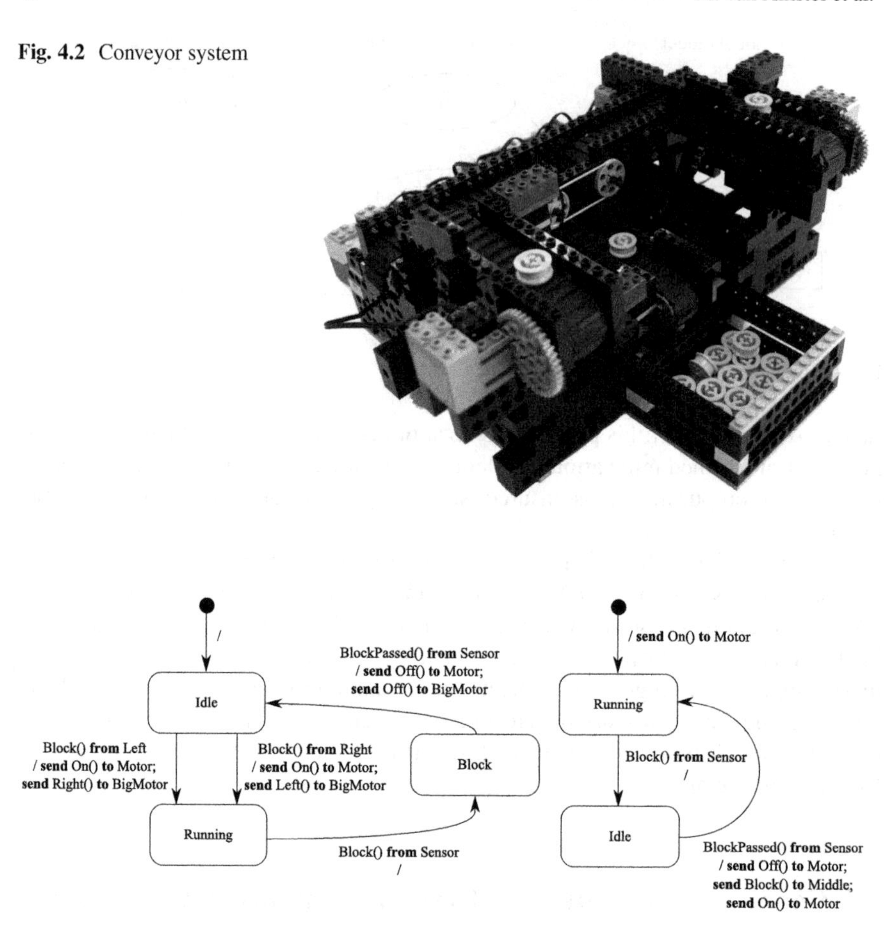

Fig. 4.3 State machines for the conveyor system depicted in Fig. 4.2

means that some constructs available in our DSL have no direct counterpart on the implementation platform. This platform, for instance, does not offer constructs such as synchronous communication or lossless channels. Instead, communication on the implementation platform is asynchronous and takes place over a lossy channel. To enable transformation from our DSL to the implementation platform, the semantic gaps between the two platforms need to be bridged [21]. To align the semantic properties of the DSL with the implementation platform, we added a number of constructs to our DSL and implemented a number of transformations that can be used to stepwise refine models. These transformations transform those model elements that are not offered by the implementation platform to constructs that are offered by the platform, while preserving the behaviour of the model. Two examples of such transformations are discussed in the remainder of this section. A final transformation transforms the resulting model to executable code.

4.2.1.1 Synchronous Communication over Asynchronous Channels

In the first transformation, synchronous signals are replaced by asynchronous signals. To ensure that the behaviour of the model remains the same, acknowledgment signals are added for synchronisation. Whenever a signal is sent, the receiving party sends an acknowledgment indicating that the signal has been received. The sending party waits until the acknowledgment has been received. In this way, synchronisation is achieved.

4.2.1.2 Lossless Communication over Lossy Channels

To ensure lossless communication over a lossy channel, an implementation of the alternating-bit protocol (ABP) is added for every class that communicates over a lossless channel in the source model. The protocol will ensure that a signal will arrive at its destination eventually, under the premise that not all signals get lost. In this way, lossless communication over a lossy channel is achieved.

4.2.2 Pros and Cons

The advantage of this model-driven approach is that executable code can be generated from a simple model without any user intervention. The small state machines depicted in Fig. 4.3 are refined and 1,170 lines of NQC code are generated to control the physical setup depicted in Fig. 4.2. The disadvantage of this approach is that errors in a model will manifest themselves as run-time errors. When software is generated for high-cost hardware, this is undesirable. Therefore, we also implemented model transformations that can generate simulation and verification code from domain-specific models. In this way, formal methods, like simulation and verification, can be used without having to create models suitable for that purpose separately. In collaborations with industrial partners, we observe that they find formal methods useful, but hard to use. Using a transformational approach such as the one presented here, formal methods can be applied without having to know their nitty gritty details.

4.3 Formalisation of Material Flow Diagrams: A Case Study

In this section, we present a second case study. It involves the formalisation of a DSL (i.e. a meta-model) for describing a material handling system (material flow diagrams). Since a typical industrial process uses many DSLs, this study can serve as an example for developing those languages.

A material flow diagram (MFD) is a model that plays a prominent role in the development process of a material handling system. An MFD contains the functional elements that are part of a material handling system, and their connections. Typically, an MFD is created as an informal drawing. Therefore, the information contained in an MFD has to be interpreted and transferred manually to other documents in different phases of the development process of a material handling system. Due to the problems mentioned in Sect. 4.1, there is a need to formalise MFDs in such a way that they can be faithfully transferred to different steps of the development process (semi-)automatically [2].

4.3.1 Development Process

The formalisation process consists of the following steps. First, a domain analysis is performed. In this phase, the concepts that play a role in the domain and their relations are determined. Second, these concepts and relations are captured in a DSL. This DSL describes the structure of an MFD, i.e. the entity types, attributes of entities, and allowed relations between the entities. Third, a graphical editor is created to graphically construct models conforming to the designed meta-model. Last, model transformations are developed to generate artefacts that can be used to analyse MFDs, as well as artefacts required in various phases of the development cycle.

4.3.2 Formal Material Flow Diagrams

The development of the DSL was an iterative process in which a number of alternatives were considered. In the end, an extensible DSL has been developed to be flexible with respect to the type of material handling system that needs to be modelled, e.g. warehouses, express parcel systems, etc. The developed DSL consists of a generic part facilitating the definition of hierarchies, and generic components and connectors. In itself, this generic part is not usable, it needs to be merged with a library. Such a library contains the components, such as workstations and conveyors, and connectors, such as physical and data connections, for a specific type of material handling system. Merging the generic DSL with a library is performed automatically by means of a model transformation.

A formal DSL ensures that the MFDs created using it are *syntactically* correct. This is, however, not enough. Therefore a set of constraints has been defined and integrated into the DSL. These constraints ensure, to a certain extent, static semantic correctness of the MFDs.

Currently, designers create MFDs using Microsoft Visio. However, the models created in this way have no formal basis, and can therefore not be used further in an automated development process. To facilitate designers, who are used to diagrammatic MFDs, in creating formal models, a graphical editor has been developed. Figure 4.4 depicts a screenshot of the graphical editor.

One of the goals of creating formal MFDs is to generate artefacts from them that can be used for analysis or for further development of a material handling system. Therefore, three model transformations have been developed. The first model transformation facilitates simple analysis. It calculates a number of metrics from an

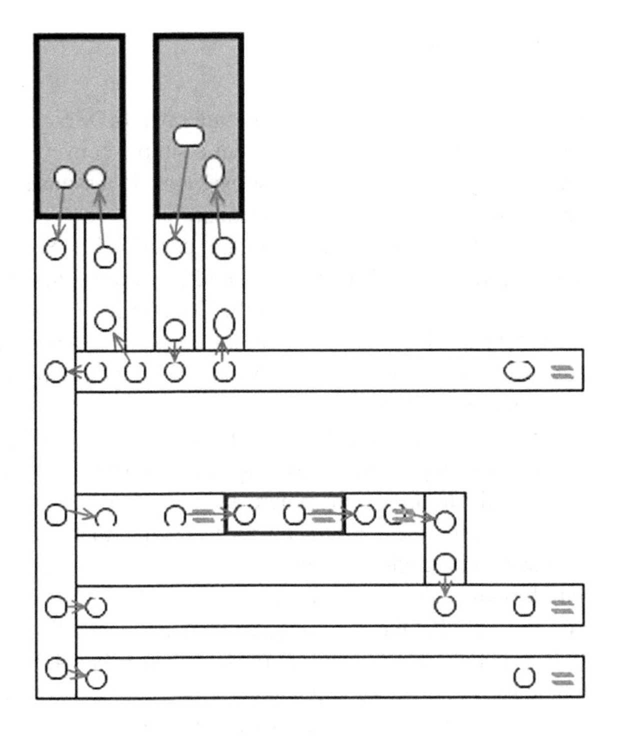

Fig. 4.4 An example MFD created using the graphical MFD editor

MFD. These metrics give insight into the complexity of an MFD. The second model transformation is aimed at generating code for system-level control, i.e. code that concerns among others planning and routing in a material handling system (see Chaps. 2 and 3). When this model transformation was being developed, the system-level control framework (presented in Chap. 2) was not yet ready. Therefore, only an architectural diagram is generated. However, adapting this transformation to generate actual code for the framework is rather straightforward. The last model transformation is used to generate input for the generic transport routing simulator described in Chap. 7. Using this simulator, a quick performance analysis of the material handling system can be performed. An MFD describes how components in a material handling system are connected. This layout information serves as the basis for the simulation. However, an MFD is an abstract description of a material handling system, and does not contain information regarding the position of the components in it. Therefore, the sizes of components such as, for example, conveyors, are derived from the diagram created using the graphical editor. Although this does not accurately represent the sizes in the final material handling system, this proof of concept shows that an MFD together with position information of its components can be used to simulate a material handling system.

4.4 Reflection

In Sects. 4.2 and 4.3, we have presented two MDSE case studies. We have used the DSL presented in Sect. 4.2 to model a simple conveyor system, consisting of only three conveyor belts. Although some valuable lessons have been learned from this, the case study is not on an industrial scale.

The MFDs created using the tools described in Sect. 4.3 have several benefits. Since the MFD models adhere to a DSL, they are syntactically correct. A number of constraints have been defined on the DSL, to ensure the static semantic correctness of the MFDs. Model transformations enable analysis of the MFD models, but also increase their portability, i.e. they can automatically be adapted for use in different stages of the development.

Since the MFD has such a prominent role in the development process, formalising them is a first step towards adopting a model-driven approach to develop warehouses. However, the tools presented here are not yet mature enough to deploy directly. The DSL containing the components for modelling material handling systems should be extended and the editor should be made more user-friendly. A next step would then be to formalise the models based on MFDs that are created later in the development process, such that they can be generated automatically for a large part.

The model transformations presented in Sects. 4.2 and 4.3 are not sufficiently mature for an industrial application. We believe, however, that the presented model-driven approach can also be applied on a larger scale. There are however different issues that will play a role. First, models need to be developed and maintained by multiple developers. To facilitate this, configuration management techniques tailored towards models need to be adopted. One of the key components of such a system deals with comparing models. We will elaborate on this in Sect. 4.5. Second, since model transformations are being employed, there will be a shift from maintaining code to maintaining models and the corresponding transformations. To prevent model transformations from becoming the next *maintenance nightmare*, their quality should be monitored and analysis techniques should be developed for them to assist in the maintenance process. We will elaborate on this in Sect. 4.6.

4.5 Model Comparison

One of the basic functions of a model configuration management system is efficient storing of models. In order to achieve this, model configuration management systems store only the initial version of a model, and the differences between the initial and any subsequent version. Another important function is providing insight into the evolution of models. In order to achieve this, model configuration management systems must visualise the differences between two models. Furthermore, model configuration management systems allow multiple developers to work on the same project concurrently. Thus, it is important to know the difference between versions of models developed by different developers. For example, if two developers develop a model of a warehouse in parallel, they would probably need to compare their models and agree on the common model. It is clear that all of these functions rely on model comparison techniques. Thus, good model comparison techniques are crucial for the success of any model configuration management system, and with it to the entire MDSE approach.

The result of the process of comparing two models is the set of model differences. Since it is safe to assume that in the engineering process more than one DSL will be used, the differences between models should not be dependent on the DSL, i.e. the differences should be DSL-independent.

The comparison process itself can be split into two sub-processes: model matching and differences calculation. The role of model matching is in determining which elements in the first compared model correspond to which elements in the second compared model. The role of differences calculation sub-process is in calculating the set of differences based on the found matchings.

In the course of the Falcon project, we developed a novel approach to model comparison [22, 23]. This approach is DSL-independent and, unlike traditional approaches, supports all traditional model matching methods. The DSL-independence is achieved by relating model differences not to DSLs, but to the language that is used to describe DSLs, a meta-DSL. All traditional matching methods are supported by designing a highly configurable matching algorithm. The configurations for the algorithm can and should be specified for each DSL used. The reason for this is that each DSL has a different intended meaning, and thus the speed and precision of a basic matching algorithm can be improved by a domain-specific configuration.

We validated our approach by building a tool, which is described in detail by Van den Brand et al. [24].

4.6 Analysis of Model Transformations

Model transformations are in many ways similar to traditional software artefacts, i.e. they have to be used by multiple developers, have to be changed according to changing requirements, and should preferably be reused. Moreover, model transformations are artefacts that will have a long lifespan and are to be maintained by several people. Therefore, it is necessary to define and assess their quality. In an industrial context this is particularly important. For example, if a developer developing a particular transformation retires, a transformation of good quality will be reusable by his successors.

4.6.1 Quality of Model Transformations

Quality attributes such as modifiability, understandability, and reusability need to be understood and defined in the context of MDSE, in particular for model transformations. For most other types of software artefacts, e.g. source code and models, there already exist approaches for measuring their quality. The goal is to make the quality of model transformations measurable. The type of quality we refer to here is the internal quality of model transformations, i.e. the quality of the transformation artefact itself [17].

An approach that is often employed for assessing the quality of software artefacts is metrics. We will use metrics for assessing the quality of model transformations as well. For four model transformations formalisms, we have defined a set of metrics, i.e. for ASF+SDF [25], ATL [7], QVT operational mappings [9], and Xtend [5]. These metric sets can be used for measuring model transformations created with each of the respective model transformation formalisms. The metrics are defined on the model transformation artefact only, i.e. the input and output models of a model transformation are not considered. This is referred to as direct assessment [17]. By means of empirical studies, we assess whether the metrics are valid predictors for a number of quality attributes [18]. Once we have identified the properties of model transformations that affect their quality, we can propose a methodology for improving that quality. This methodology will probably consist of a set of guidelines which, if adhered to, lead to high-quality model transformations.

4.6.2 Visualisation Approaches for Model Transformations

Numerous analysis techniques supporting software maintenance exist for all kinds of software artefacts, such as source code or models. However, few techniques currently exist for analysing model transformations. A reason for this is that MDSE and thereby model transformations is a relatively young research discipline and most effort is invested in applications and in improving model transformation techniques and tools (see the proceedings of the 2008, 2009, 2010 ICMT conferences [16, 10, 15]). To prevent model transformations from becoming the next *maintenance nightmare*, analysis techniques should be developed for them to assist in the maintenance process. Moreover, proper tool support is required for further adoption of MDSE in industry [8].

A significant proportion of the time required for maintenance, debugging, and reusing tasks is spent on understanding [13]. Therefore, we developed two techniques facilitating model transformation comprehension.

4.6.2.1 Call Relation Visualisation

The first analysis technique we developed, makes the call relation between transformation functions visible, thus revealing dependencies between various transformation functions. For the visualisation of these call relations, we use the tool ExtraVis [4]. In Fig. 4.5, the call relation of one of the transformations used in the case presented in Sect. 4.2 is depicted. Visualisation of dependency data has various useful

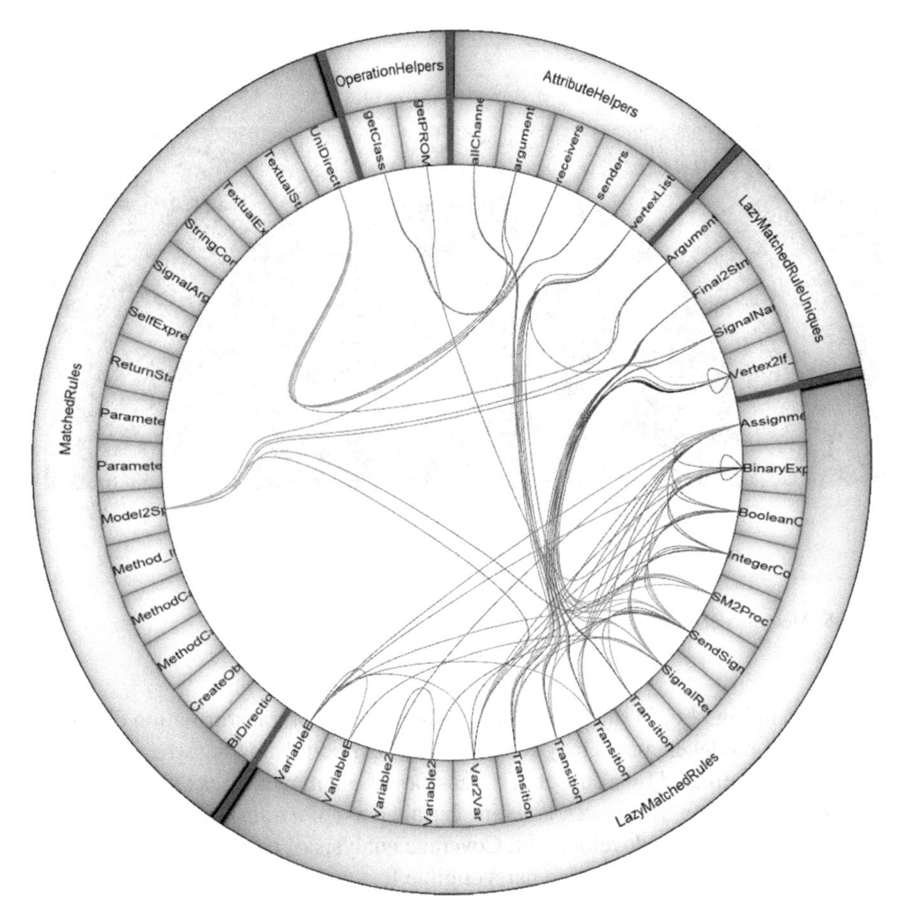

Fig. 4.5 Trace visualisation. The outer ring displays the modules that comprise the transformation. The second ring displays the different kinds of transformation function types per module

applications in the development and maintenance process of a model transformation. By actually seeing in what way transformation functions depend on each other and interact, the understanding of the transformation may increase.

Moreover, parts of the transformation that are obsolete can be identified and subsequently be removed. Furthermore, this visualisation can be used to reveal code smells in transformations, like modules with low cohesion and high coupling [12].

4.6.2.2 Meta-Model Coverage Visualisation

The second analysis technique we developed, provides insight in the part of the DSL that is covered by a model transformation. For this purpose we developed two visualisation techniques for coverage analysis. An example of one of these visualisations

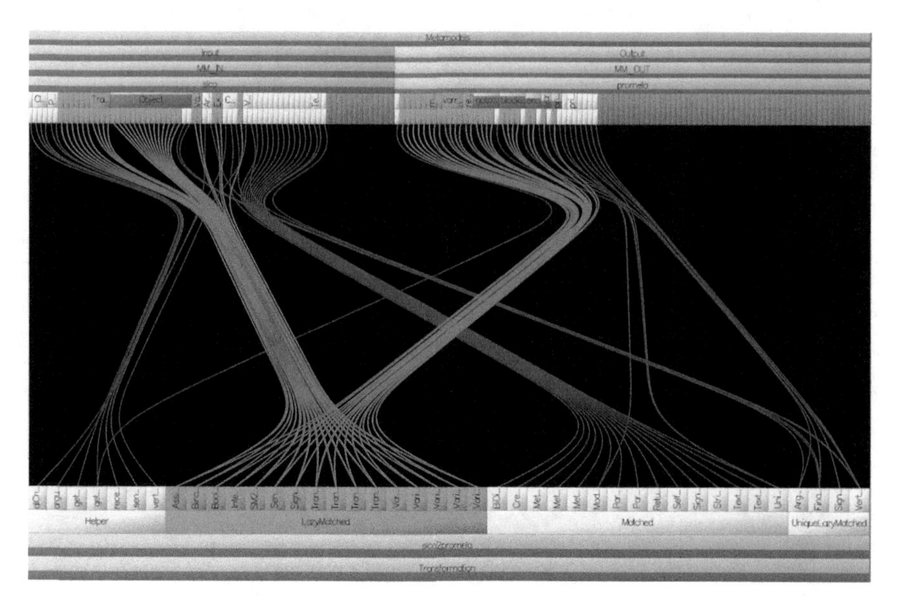

Fig. 4.6 Meta-model coverage relation visualisation

is depicted in Fig. 4.6. In this figure, the relations between transformation elements and the meta-model elements they cover are made explicit. Knowing what part of the meta-model is affected by a subset of a transformation, may increase understanding of the transformation. Moreover, isolating certain parts of a transformation may help in finding errors during development. Coverage analysis of part of a transformation may also be useful for identifying parts eligible for reuse.

4.7 Conclusion

In this chapter, we explained MDSE, a software engineering discipline aimed at dealing with the increasing complexity of software. We presented two of our case studies, demonstrating the application of MDSE.

To reap the full advantages of MDSE, much research and development is still needed. In this chapter, we presented the results of our research aimed at improving MDSE processes. First, we discussed a methodology for comparing models, which is a pivotal component for model configuration management. We validated our approach by comparing a large number of models, using a tool we developed. Second, we explained the necessity for assessing the quality of model transformations. We also presented two visualisation approaches that can help increase the understanding of model transformations.

References

1. Baum D, Hansen J (2003) NQC programmer's guide. http://bricxcc.sourceforge.net/nqc/doc/NQC_Guide.pdf. Viewed May 2011
2. Bijl RJ (2010) Formalizing material flow diagrams. Master's thesis, Department of Mathematics and Computer Science, Eindhoven University of Technology, Eindhoven
3. Brooks FP Jr (1995) The mythical man-month: essays on software engineering, anniversary. (2nd) edn. Addison-Wesley Professional, Boston
4. Cornelissen B, Holten D, Zaidman A, Moonen L, van Wijk JJ, van Deursen A (2007) Understanding execution traces using massive sequence and circular bundle views. In: 15th IEEE international conference on Program comprehension, ICPC '07 pp 49–58
5. Haase A, Völter M, Efftinge S, Kolb B (2007) Introduction to open Architecture Ware 4.1.2. In: Model-driven development tool implementers forum (MDD-TIF'07) (co-located with TOOLS 2007)
6. Jones C (2004) Software project management practices—failure versus success, CrossTalk. J Def Softw Eng 17:5–9
7. Jouault F, Kurtev I (2005) Transforming models with ATL. In: Model transformations in pactice workshop, Lecture notes in computer science, vol 3844. Springer, Berlin, pp 128–138
8. Mohagheghi P, Dehlen V (2008) Where is the proof?—A review of experiences from applying MDE in industry. In: Model driven architecture foundations and applications. Lecture notes in computer science, vol 5095. Springer, Berlin, pp 432–443
9. Object Management Group (2008) Meta Object Facility (MOF) 2.0 Query/View/Transformation specification, version 1.0. Document formal/2008-04-03, OMG
10. Paige RF (ed) (2009) Theory and practice of model transformations. In: Proceedings of 2nd international conference, ICMT 2009. Lecture notes in computer science, vol 5563. Springer, Berlin, Zurich 29–30 June, 2009
11. Schmidt DC (2006) Model-driven engineering. IEEE Comput 39:25–31
12. Stevens WP, Myers GJ, Constantine LL (1974) Constantine LL Structured design. IBM Syst. J 13:115–139
13. Storey MAD, Wong K, Muller HA (2000) How do program understanding tools affect how programmers understand programs? Sci. of Comp. Program 36:183–207
14. The Standish Group (1995) The Standish Group report. http://www.standishgroup.com/chaos/intro1.php. Viewed November 2010
15. Tratt L, Gogolla M (eds.) (2010) Theory and practice of Model Transformations. In: Proceedings of 3rd international conference, ICMT 2010. Lecture notes in computer science, vol 6142. Springer, Berlin, Malaga 28 June–2 July, 2010
16. Vallecillo A, Gray J, Pierantonio A (eds) (2008) Theory and practice of model transformations. In: Proceedings of 1st international conference, ICMT 2008. Lecture notes in computer science, vol 5063. Springer, Berlin, Zürich 1–2 July, 2008
17. van Amstel MF (2010) The right tool for the right job: assessing model transformation quality. In: Computer software and applications conference workshops (COMPSACW), IEEE 34th annual, pp 69–74
18. van Amstel MF, Lange CFJ, van den Brand MGJ (2009) Using metrics for assessing the quality of ASF+SDF model transformations. In: Theory and practice of model transformations, Lecture notes in computer science, vol 5563. Springer, Berlin, pp 239–248
19. van Amstel MF, van den Brand MGJ, Engelen LJP (2010) An exercise in iterative domain-specific language design. In: Proceedings of the joint ERCIM workshop on software evolution (EVOL) and international workshop on principles of software evolution (IWPSE), pp 48–57
20. van Amstel MF, van den Brand MGJ, Engelen LJP (2011) Using a DSL and fine-grained model transformations to explore the boundaries of model verification. In: Proceedings of the 3rd workshop on model-based verification & validation from research to practice (MVV 2011)

21. van Amstel MF, van den Brand MGJ, Protić Z, Verhoeff T (2008) Transforming process algebra models into UML state machines: bridging a semantic gap? In: Theory and practice of model transformations. Lecture notes in computer science, vol 5063. Springer, Berlin, pp 1–75
22. van den Brand MGJ, Protić Z, Verhoeff T (2010) Fine-grained metamodel-assisted model comparison. In: Proceedings of the 1st international workshop on model comparison in practice, pp 11–20
23. van den Brand MGJ, Protić Z, Verhoeff T (2010) Generic tool for visualization of model differences. In: Proceedings of the 1st international workshop on model comparison in practice, pp 66–75
24. van den Brand, MGJ, Protić Z, Verhoeff T (2010) RCVDiff - a stand-alone tool for representation, calculation and visualization of model differences. In: Proceedings of the international workshop on models and evolution-ME 2010
25. van Deursen A (1996) An overview of ASF+SDF. In: Language prototyping: an algebraic specification approach, AMAST series in computing, vol 5, Chap.1. World Scientific Publishing, Singapore, 1–29
26. van Deursen A, Klint P, Visser J (2000) Domain-specific languages: an annotated bibliography. SIGPLAN Not. 35:26–36

Part III
Models in System Design

Chapter 5
Aggregate Models of Order-Picking Workstations

Ricky Andriansyah, Pascal Etman, and Jacobus Rooda

Abstract An aggregate model of an order-picking workstation is proposed for an automated warehouse with a goods-to-man order-picking system. A key aspect of the model is that the various stochasticities that contribute to the workstation performance are not modelled in detail. Rather, these are aggregated into a single effective process time distribution, which is calculated from arrival and departure times of products at the workstation. Two model variants have been developed namely for workstations processing products and orders in first-come-first-serve (FCFS) and non-FCFS sequence. Both model variants have been validated using data obtained from an operating order-picking workstation. The use of aggregate models in practice is then elaborated.

5.1 Introduction

Developing a new warehouse automation concept often means pushing the boundaries of feasibility of material handling technology. Even when this is successfully achieved, market uncertainty still exists regarding the customer acceptance of the new concept. Moreover, developing and realising the new concept must be done quickly to achieve a short time-to-market. All of these typically cause a high development cost for such a complex system. With this regard, models that allow quick adjustments

R. Andriansyah (✉) · P. Etman · J. Rooda
Department of Mechanical Engineering, Eindhoven University of Technology,
P.O. Box 513, 5600 MB Eindhoven, The Netherlands
e-mail: r.andriansyah@tue.nl

P. Etman
e-mail: l.f.p.etman@tue.nl

J. Rooda
e-mail: j.e.rooda@tue.nl

R. Hamberg and J. Verriet (eds.), *Automation in Warehouse Development*,
DOI: 10.1007/978-0-85729-968-0_5,

and analysis while still giving satisfactory accuracy to reality are crucial to provide insights on system performance early in the design phase. Such models can be used to compare the added value from the new concept relative to the development cost, which eventually supports decision-making processes.

Robustness of system performance is also a relevant issue in developing highly automated warehouses. For such systems with large investment of money and resources, it is no longer sufficient to have a working system under a predefined, specific setting. The next generation warehouses are "flexible platforms" that provide technically feasible stepping stones for further development to anticipate changes in the future. That is, the warehouses are robust in performing under continuously changing customer requirements. Reusability, exchangeability, and flexibility are the keywords for highly automated warehouses.

This chapter addresses a goods-to-man order-picking system (OPS) with remotely located order-picking workstations. Such a system is characterised by a number of stochastic behaviours. Conventional performance analysis techniques necessitate parametrisation of each factor contributing to the workstation throughput performance. However, only incomplete information is typically available regarding these factors, if not based on assumptions. Moreover, the available information may only give expected rather than effective values.

With this regard, the main question is how to develop an accurate performance analysis model based on typically limited available logging data. An aggregate modelling technique based on the concept of effective process time is proposed. A key aspect of the aggregate model is that the various stochasticities that contribute to the flow time performance are not modelled in detail. In practice, these are typically difficult to quantify [12]. Instead, they are modelled into a single aggregate process time distribution. Our goal is to obtain the aggregate process time distribution from product tote arrival and departure events of the order-picking workstation in operation. Such an aggregate model can be used to quantify the product and order flow time distribution of an order-picking workstation.

5.2 System Description

The goods-to-man OPS discussed here comprises three separate units, namely *miniloads*, *workstations*, and *a closed-loop conveyor*. Miniloads provide temporary storage spaces for *product totes*. A product tote contains a number of items of a particular stock-keeping unit (SKU), which may be required by an order. At the workstation, items are picked from product totes and put into order totes to fulfil customer orders. A closed-loop conveyor connects the miniloads to the workstations. This OPS is shown in Fig. 5.1.

Miniloads are automated storage racks equipped with cranes to serve two functions, namely storing and retrieving product totes. Each miniload consists of two single-deep racks with a single storage/retrieval crane in the middle to access

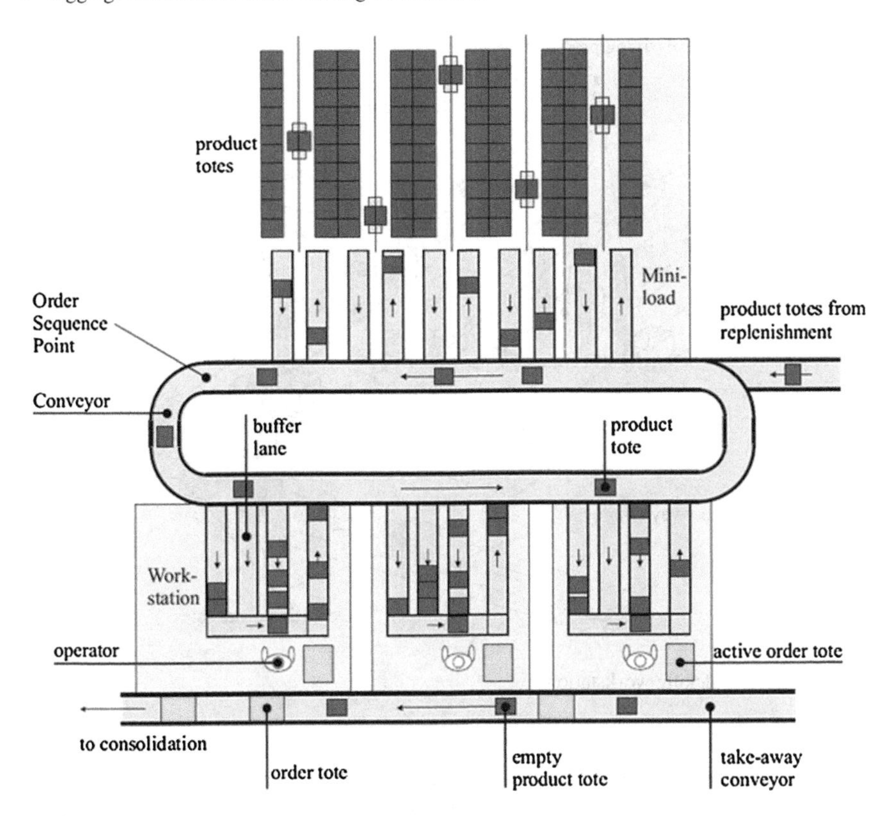

Fig. 5.1 A goods-to-man order-picking system

product totes. The cranes move horizontally along the aisle between the racks, while a holder of product totes moves vertically to store or retrieve the totes.

The closed-loop conveyor transports product totes from the miniloads to the workstations, and vice versa. As there is only limited space on the conveyor, only product totes that have successfully reserved space on the conveyor are allowed to enter the conveyor.

An order-picking workstation is shown in Fig. 5.2. At such a workstation, a picker works to fulfil *orders*. An order consists of a number of *order lines*. The number of order lines in an order is referred to as *order size*. Order sizes may vary significantly. Internet orders, for example, may have a small order size while orders from supermarkets may have a very large order size. An order line represents the required number of *items* from a certain SKU.

The following processes take place at an order-picking workstation. Product totes arrive at the workstation and form queues on the buffer conveyors. Once the picker and the required product tote are available, the product tote will be removed from the buffer conveyor and transported to the pick position where the picker stands. The picker then picks a number of required items from the product tote and puts them in an *order tote*. The picker works on one order at a time until all lines of the

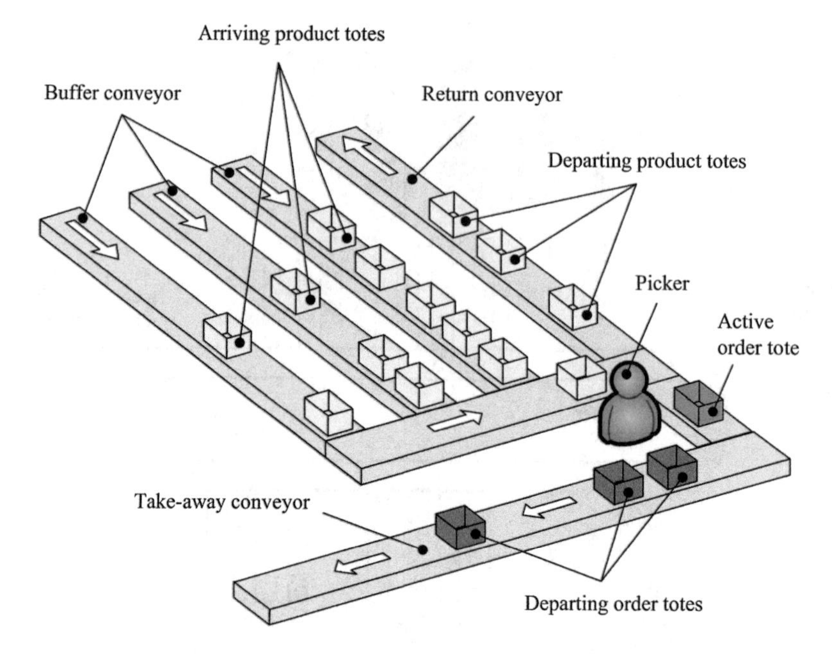

Fig. 5.2 An order-picking workstation

order have been picked and the order is said to be finished. When an order is finished, the picker moves the finished order tote to a take-away conveyor that brings the order tote to a consolidation area. An order tote corresponds to one order. If a product tote is not yet empty after item picking, the tote will be returned to the storage area.

Order-picking workstations have a typical characteristic that distinguishes them from ordinary manufacturing workstations. An order-picking workstation receives a number of product totes for different orders. In the type of system that we consider, the picker can only pick items from the product totes that belong to the order currently being processed, known as the *active order*. The product totes required to fulfil the active order are referred to as *active totes*. As such, only active totes are sent from the buffer conveyor to the pick position, while all other product totes wait in the queue. If there are no active totes in the queue, then the picker will be idle although totes for other orders may be present.

Two performance measures are particularly of interest for this order-picking workstation, namely the tote and order flow times. Tote flow time is defined as the total time spent by a product tote at the order-picking workstation, which starts when a product tote arrives at the workstation and ends when it departs the workstation. Order flow time is defined as the time required to complete an order, which starts when the first product tote of an order arrives at the workstation and ends when the last product tote of the order has left the workstation. A complete order means that all items required for the order have been picked into the corresponding order tote.

5.3 Models of Goods-to-Man Order-Picking Systems

Two well-known model classes in the literature are analytical models and simulation models. The type of questions that can be answered using these approaches is different, and thus they are generally used at different phases for different purposes.

Analytical models are particularly useful during the system design phase. During this phase, one is mainly interested in having a quick overview of the performance of different designs. Analytical models serve this purpose under necessary assumptions for mathematical tractability. This being said, analytical models may not capture all details of the system. Most analytical performance analysis of goods-to-man OPS are based on closed queueing network models.

Simulation models are widely used alternatives to analytical models for performance analysis at the system utilisation phase. They are especially useful in evaluating what-if scenarios of (detailed) operational policies or parameters on a specific design. They also allow more details of the system to be captured than in analytical models. As such, these models are practical in identifying specific operational settings that improve system performance. However, building a valid, credible, sufficiently detailed simulation model and generating appropriate outputs may be very time consuming. Moreover, detailed simulation models may require extensive computational capability. Nevertheless, simulation is still the most widely used technique for warehouse performance analysis in practice [7].

5.3.1 The Concept of Aggregation

Regardless of the type of model used, data availability is the key for prediction accuracy in performance analysis. The absence of measurable data compromises reasonable inputs required for the model. In turn, performance analysis based on misleading assumptions is harmful for decision making. It is therefore crucial that all model parameters can be limited to data that is obtainable from the shop-floor.

Unfortunately, data collection in such a complex system is not straightforward. A goods-to-man OPS exhibits various stochastic behaviours. In addition to inter-arrival time and picking time, stochasticities due to interruptions are typically difficult to quantify, if even possible. To overcome this problem, an aggregate modelling technique based on effective process time (EPT) is proposed.

The concept of process time aggregation is as follows. An order-picking workstation is characterised by several process time components (see Fig. 5.3). At the core of the process is the time required for picking items, which is referred to as the raw pick time. In addition to the raw pick time, pickers may require some setup time (change-over time) between processing of orders. Conveyor systems may break down, causing unavoidable delays. Picker availability is also an issue, since it is likely that a picker is sometimes not present at the workstation. In an *aggregate model*, these components are aggregated into a single EPT distribution. The idea is then to reconstruct the distribution directly from tote arrival and departure times

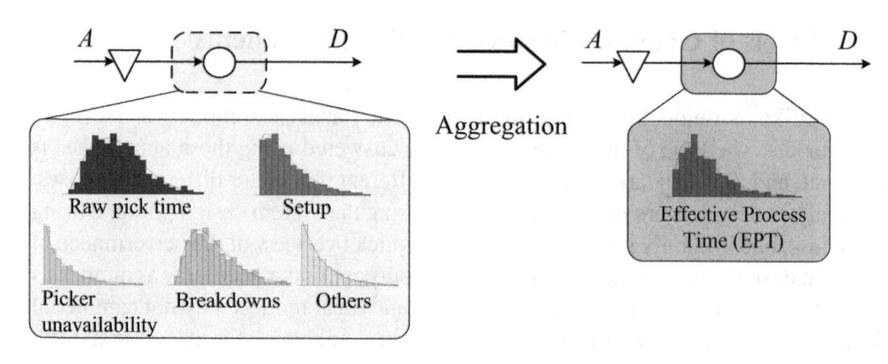

Fig. 5.3 Aggregation method

that are typically available from the logging data. An obvious advantage is that one does not need to quantify each component contributing to the process time. A similar approach to that of Kock et al. [8–10] is used to calculate the EPTs namely using a sample path equation.

5.4 Effective Process Time-Based Aggregate Models

Aggregate models use EPT realisations as input. An EPT realisation is calculated for each departing tote, which equals the total amount of time a tote *claims* capacity even if the tote is not yet in physical process. When EPT realisations for all departing totes have been obtained, an EPT distribution with mean t_{e} and squared coefficient of variation c_{e}^2 is created. We typically assume a gamma distribution, but other distributions may equally well be used. A gamma distribution is relatively easy to construct since the scale and shape parameters are readily obtainable from the mean and variance of the empirical EPT realisations.

Figure 5.4 shows an example of arrivals and departures of six totes at an order-picking workstation. Totes 1, 2, and 3 belong to order p, tote 4 belongs to order q, and totes 5 and 6 belong to order r. Arrival A_i occurs at the moment a product tote i enters the buffer conveyor of the order-picking workstation. Departure D_i occurs when item picking has been finished and the respective product tote i is moved to the return conveyor or to the take-away conveyor (see Fig. 5.2).

EPT realisations are calculated using the following sample path equation:

$$\mathrm{EPT}_i = D_i - \max\{A_i, D_{i-1}\}, \tag{5.1}$$

where D_i denotes the time epoch of ith departing tote. A_i denotes the arrival epoch of the corresponding ith departing tote. The bottom part of Fig. 5.4 illustrates how EPT realisations are obtained using Eq. 5.1.

Two aggregate model variants have been developed, namely for workstations with first-come-first-serve (FCFS) [5] and for workstations with non-FCFS processing

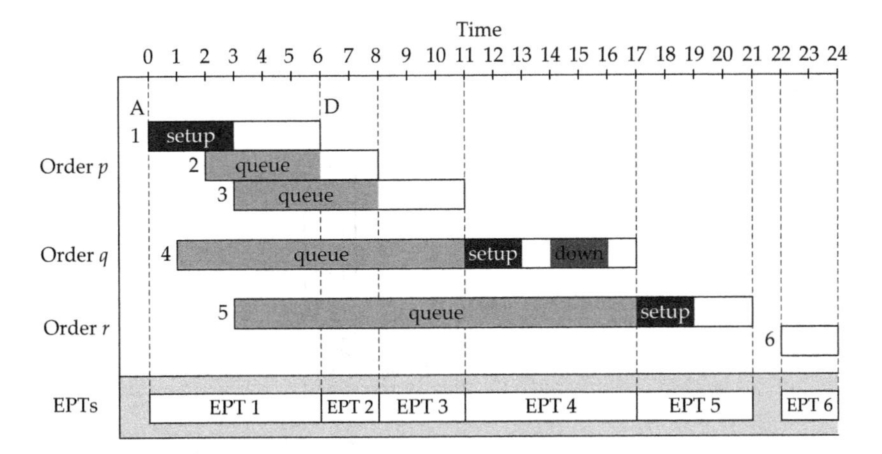

Fig. 5.4 Gantt chart example

sequence of products and orders [4]. The EPT distributions are used as input for both model variants.

5.4.1 First-Come-First-Serve processing

An aggregate model with FCFS processing as shown in Fig. 5.5 is proposed. It is a *polling system* with a single server S and k infinite queues. A number of queues is available as denoted by $Q_i, i = 0, \ldots, k-1$, where k indicates the maximum number of orders for which product totes *simultaneously* arrive in the workstation. In Fig. 5.5 we assume that product totes for three orders may arrive simultaneously, hence $k = 3$. Each tote has an *id* that denotes the order ID to which the tote belongs. All arriving totes with the same *id* are put into the same queue.

When the first tote of a new order enters a queue, a *gate* is immediately set for that queue. The gate indicates the number of totes required for the order, which equals to the order size. The gate is kept open until all totes for the corresponding order have arrived at the queue. Once the last tote of the order has arrived, the gate is closed. In Fig. 5.5 an open gate is represented by a dotted line and a closed gate is represented by a solid line in the queue. A new order is created each time the gate for another order has been closed. The variable *id* is increased by one and the totes arriving for new order are put in queue Q_i where $i = id$ modulo k.

The server attends the queues in a cyclic direction, causing the orders to be served in a FCFS sequence. The server will switch to the next queue only if the gate for the current queue has been closed and all totes in front of the gate have been served. If the server is done processing all totes in front of the gate, but the gate is still open, then the server will become idle at the queue. In this case, the server waits until the remaining totes for the queue arrive.

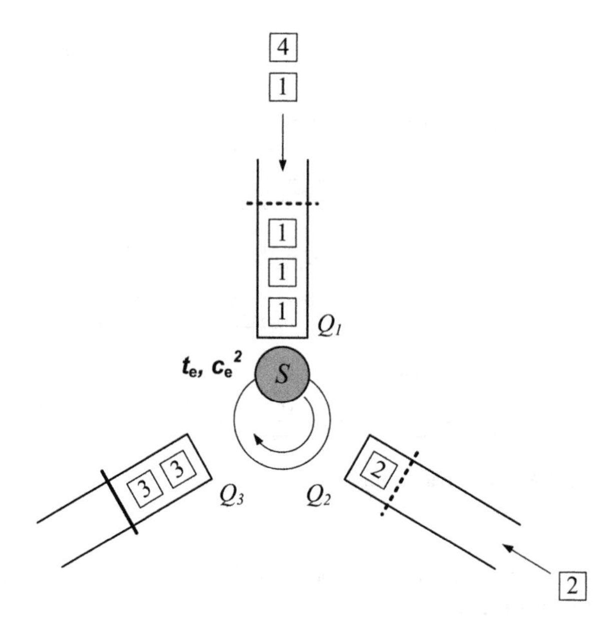

Fig. 5.5 Aggregate model of an order-picking workstation with FCFS processing

The processing time of the server in the aggregate model of Fig. 5.5 is sampled from an EPT distribution with mean t_e and squared coefficient of variation c_e^2. The EPT distribution is the only input used in the aggregate model.

5.4.2 Non-First-Come-First-Serve Processing

Under the assumption of non-FCFS processing, tote and order overtaking may occur at this workstation. Overtaking takes place when the next tote/order processed by the picker is not the oldest active tote/order in the buffer.

An aggregate model as shown in Fig. 5.6 is proposed. It is essentially a single-server queueing system with an infinite buffer. Within the buffer there are a number of infinite queues, each containing totes for an order. Three inputs are used in this aggregate model, namely EPT distributions, overtaking distributions, and decision probabilities. These inputs are calculated based on data of arrival and departure times of totes. When the server is idle, an active tote in the buffer may be processed based on a so-called decision probability and overtaking distribution. The decision probability gives the probability that an active tote will be processed or not by the idle server. The overtaking distribution is used to determine which active tote in the buffer will be selected as the next tote to be processed. The server then processes the selected active tote with a processing time sampled from the EPT distribution.

Figure 5.6 shows an example of both tote and order overtaking. The server is currently processing tote 2.2, which is the second arriving tote of order 2. However, order 1 arrived earlier than order 2 (that is, order 1 is positioned in front of order 2). Hence, order 2 overtakes order 1. Also, the first arriving tote of order 2 (tote 2.1) has

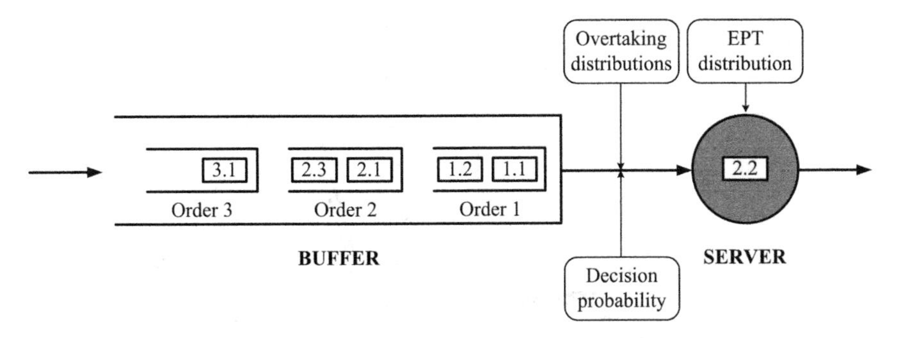

Fig. 5.6 Aggregate model of an order-picking workstation with non-FCFS processing

not been processed yet. As such, the second tote of order 2 (tote 2.2) overtakes the first tote of order 2 (tote 2.1).

The three inputs for the aggregate model are measured directly from tote arrival and departure data of an operating order-picking workstation. Using the inputs, the aggregate model predicts the mean and variability of tote and order flow times of the operating order-picking workstation.

5.4.3 Validation Using Real Data

For both aggregate model variants we provide a case study to validate the EPT-based aggregate modelling technique [4, 5]. Using product arrival and departure times from logging data of an operating order-picking workstation, we calculate the EPT realisations and subsequently construct the EPT distribution. The EPT distribution is used as input in the aggregate model. The aggregate model is then used to generate flow time distributions of totes and orders.

Figure 5.7 shows an example of comparison of the tote flow time distribution (φ_{tote}) and order flow time distribution (φ_{order}) from the real data and the aggregate model with non-FCFS processing. The figure suggests that the EPT-based aggregate modelling technique gives a good accuracy in predicting tote and order flow time distributions.

5.5 The Use of Aggregate Models

Now that the aggregate model has been validated, it can be used for different purposes. One of the most important use is to analyse the effect of different settings, e.g. inter-arrival rates, order size distributions, and order release strategies, on the system throughput and flow times. As an example, Fig. 5.8 shows the predicted performance

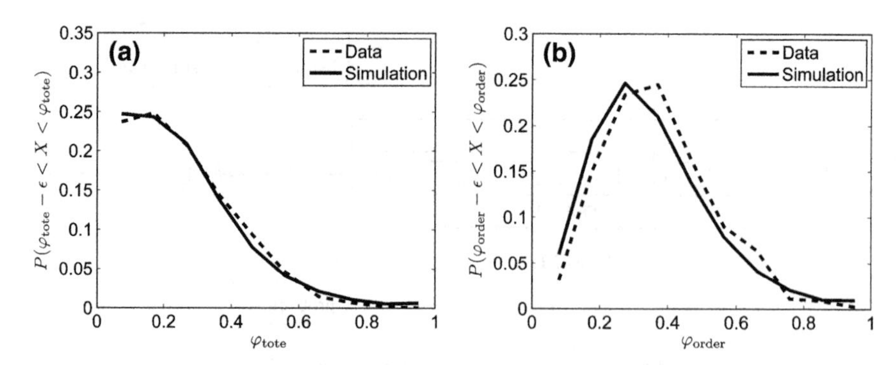

Fig. 5.7 An example of predicted flow time distributions using EPT-based aggregate model with non-FCFS processing

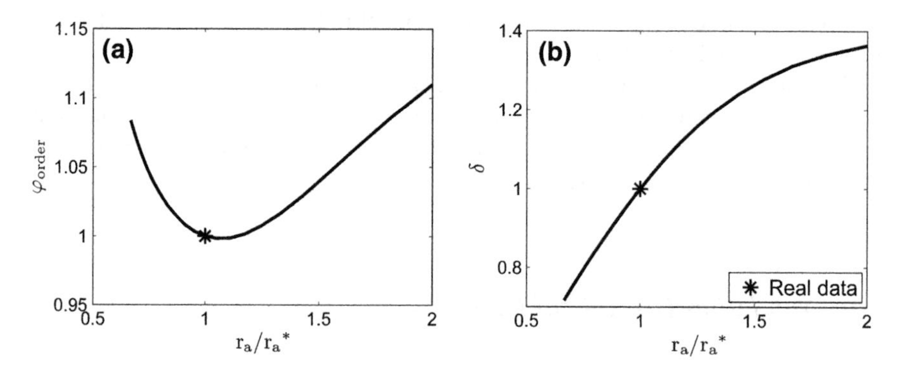

Fig. 5.8 Predicted performance under various inter-arrival rates

of an order-picking workstation under various inter-arrival rates r_a by an aggregate model with FCFS processing. In this figure, r_a^* denotes the inter-arrival rate of totes at the workstation as measured from the data of an operating order-picking workstation. Figure 5.8a and b show the mean order flow time φ_{order} and order throughput δ, respectively, as a result of changing the inter-arrival rate of totes at the workstation. The figures show normalised values.

Furthermore, EPTs can be calculated in real-time to monitor the effective pick performance of an operating order-picking workstation. Recall that EPT realisations are calculated using a simple equation with as input arrival and departure times of totes obtainable from logging data. Once EPTs are calculated, they can be visualised on the shop-floor level. This way, any deviations from the expected performance (e.g. extremely large EPTs) can be detected and the necessary corrective actions can be performed timely.

The aggregate model can be integrated into a detailed simulation model for OPS design/redesign purposes. A flexible model architecture shown in Fig. 5.9 has been proposed [3] that supports the integration. Incorporating an aggregate model into the architecture for design/redesign purposes is advantageous because data measured

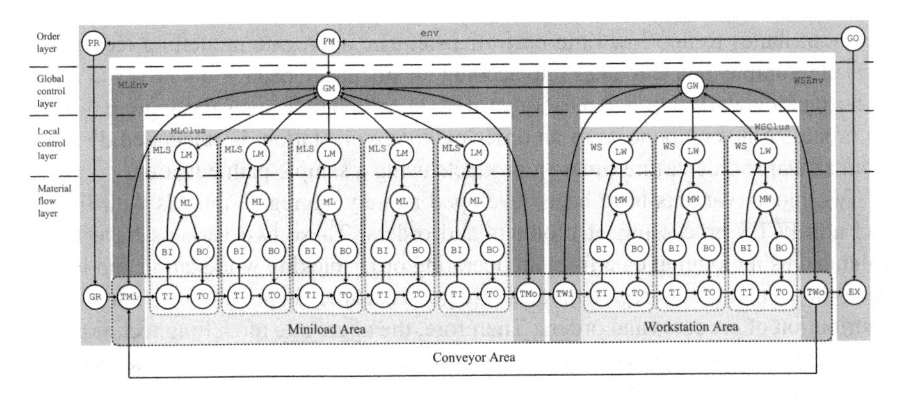

Fig. 5.9 Architecture of goods-to-man order-picking system

directly from an operating order-picking workstation is used. That is, the aggregate model is used as a black-box approach that gives the effective rather than the expected performance of a workstation.

The proposed architecture provides a general framework for performance analysis of goods-to-man OPS. Using this architecture, it is possible to evaluate the throughput of the whole system or some parts of the system. The architecture has a decentralised control structure that comprises two layers, namely the high-level and the low-level control layers. Such a control structure allows an easy implementation of various control heuristics. That is, the architecture is reusable for investigating what-if scenarios regarding e.g. system configuration, storage assignment, retrieval sequencing, and order release strategies to anticipate changes to warehouse requirements. Note that the proposed architecture has a comparable structure to the warehouse management and control system (WMCS) reference architecture described in Chap. 2. A detailed description of each component in this architecture can be found in [2].

As a final note, the EPT distribution calculated using the sample path equation can also be used for an analytical queueing network model. Having the EPT distribution as the only input simplifies the model and thus improves mathematical tractability. Such an analytical model is particularly useful to have a quick overview on the performance of different designs. Multi-class closed queueing networks models for goods-to-man order-picking systems have been developed and analysed in [6, 11, 13]. An efficient and accurate (dis)aggregation technique is developed in [11] to estimate throughput and workstation utilisations. Approximations based on mean value analysis [1] are proposed in [6, 13].

5.6 Conclusion

An aggregate modelling technique based on the concept of effective process time has been proposed for order-picking workstations in a goods-to-man OPS. The key aspect of an aggregate model is that it does not require quantification of each stochasticity that contributes to the flow time performance. The aggregate modelling technique uses measurable data and takes into account all stochastic behaviours involved at an order-picking workstation. An EPT distribution, which represents the aggregation of all process time components of an order-picking workstation, is calculated directly from arrival and departure times of products using a sample path equation.

Two model variants have been developed namely aggregate models with FCFS and non-FCFS processing of products and orders. These two variants have been validated using logging data from an operating order-picking workstation and it has been shown that the aggregate models give good accuracy in predicting the flow time distribution of products and orders. Therefore, the aggregate modelling technique is an appealing alternative for accurate performance prediction for stochastic systems under limited data availability.

Aggregate models are useful for design/redesign purposes. With this regard, integration with a detailed simulation model may be required to evaluate the overall throughput performance of a goods-to-man OPS. A model architecture has been proposed that supports integration of aggregate models of order-picking workstations with other components of a goods-to-man OPS. Such an architecture is reusable to investigate what-if scenarios regarding e.g. system configuration, storage assignment, retrieval sequencing, and order release strategies to anticipate changes to warehouse requirements.

References

1. Adan IJBF, van der Wal J (2011) Mean value techniques. In: Queueing networks: a fundamental approach, International series in operations research and management science, vol 154. Springer, Berlin, pp 561–586
2. Andriansyah R, de Koning WWH, Jordan RME, Etman LFP, Rooda JE (2008) Simulation study of miniload-workstation order picking system SE-Report 2008-07. Eindhoven University of Technology, Department of Mechanical Engineering, Eindhoven
3. Andriansyah R, de Koning WWH, Jordan RME, Etman LFP, Rooda JE (2011) A process algebra based simulation model of a miniload workstation order picking system. Comput Ind 62:292–300
4. Andriansyah R, Etman LFP, Rooda JE (2010) Aggregate modeling for flow time prediction of an end-of-aisle order picking workstation with overtaking. In: Winter simulation conference (WSC), Proceedings of the 2010, pp 2070–2081
5. Andriansyah R, Etman LFP, Rooda JE (2010) Flow time prediction for a single-server order picking workstation using aggregate process times. Int J Adv Syst Meas 3:35–47
6. Febrianie B (2011) Queueing models for compact picking systems. Master's thesis. Eindhoven University of Technology, Department of Mathematics and Computer Science, Eindhoven
7. Gu J, Goetschalckx M, McGinnis LF (2010) Research on warehouse design and performance evaluation: a comprehensive review. Europ J Oper Res 203:539–549
8. Kock AAA (2008) Effective process times for aggregate modeling of manufacturing systems. Ph.D. thesis. Eindhoven University of Technology, Eindhoven
9. Kock AAA, Etman LFP, Rooda JE (2008) Effective process times for multi-server flowlines with finite buffers. IIE Trans 40:177–186

10. Kock AAA, Wullems FJJ, Etman LFP, Adan IJBF, Nijsse F, Rooda JE (2008) Performance measurement and lumped parameter modeling of single server flow lines subject to blocking: an effective process time approach. Comput Ind Eng 54:866–878
11. Liu L, Adan IJBF (2011) Queueing network analysis of compact picking systems. Working paper
12. Rouwenhorst B, Reuter B, Stockrahm V, van Houtum GJ, Mantel RJ, Zijm WHM (2000) Warehouse design and control: framework and literature review. Europ J Oper Res 122: 515–533
13. Sun T (2010) Comparison and improvements of compact picking system models Master's thesis. Eindhoven University of Technology, Department of Mathematics and Computer Science, Eindhoven

Chapter 6
Model Support for New Warehouse Concept Development

Roelof Hamberg

Abstract Models can be used to make insights and expectations explicit in the development of new warehouse concepts. In early phases of development, models with large simplifications and assumptions can provide relevant input for taking design decisions in an economic way. This claim is illustrated in the context of developing the automated case picking concept in Vanderlande. Two critical system aspects are modelled within days while the results were instrumental in providing directions for the development team. Essential ingredients for the success of such an approach are the focus on sensitivity analysis rather than absolute figures and the credibility of the simplifications made in modelling.

6.1 Introduction

Just like with any development of a new system, the development of a new warehouse concept requires frequent comparisons of different architecture, design, and realisation options. Many decisions have to be made, often without having all relevant information at hand, and even worse, it appears that the decisions with the highest impact have to be taken at the moment when the degree of uncertainty is also highest.

In principle, different options that appear on the scene when developing a new warehouse concept could be compared by just building them. This is not very effective, as it is too expensive and it takes too long to build them all. Neither is it necessary to do so, as developers can use their knowledge and experience to compare different options, although these insights have been built up in other contexts.

A complete warehouse system consists of a complex interplay between many different components that cooperate with each other. The properties of a resulting system, when a specific set of components is put together, are difficult to predict.

R. Hamberg (✉)
Embedded Systems Institute, P.O. Box 513, 5600 MB Eindhoven, The Netherlands
e-mail: roelof.hamberg@esi.nl

R. Hamberg and J. Verriet (eds.), *Automation in Warehouse Development*,
DOI: 10.1007/978-0-85729-968-0_6, © Springer-Verlag London Limited 2012

Especially when new components or new methods of operation are involved, assumptions and reasoning are required to come up with such predictions. Moreover, the available effort to estimate any arbitrary property of every system is very limited. This chapter intends to illustrate how simple system-level models can help to address this challenge.

A few hypotheses about the use of modelling in new warehouse concept development are given here. First, it is believed that the approach is feasible, i.e. simple, quantitative system models can be made in an economic way to answer relevant questions during new concept development. Second, when analysing these models, their relative results are thought to provide more insight than the absolute figures. The effect of the design parameters on the system aspect under study yields insight about the system as a whole. The absolute values of system aspects might be estimated in a less accurate fashion, while still the effect of certain design choices can be observed unambiguously.

6.2 Development of an Automated Case Picking System

The main line of reasoning in this chapter is based on the experiences of researchers that were involved on a daily basis in the industrial development process of a logistic system, i.e. the automated case picking system, or short, automated case picking (ACP).

6.2.1 Introduction of ACP

The ACP system concept is developed by Vanderlande to automatically handle reception, storage, retrieval, and delivery of cases in distribution centres. Cases are defined to be box-shaped[1] items which are at least 15 cm and at most 60 cm in any direction. The need for an automated solution in this field is mainly driven by the fact that human pickers are expensive and difficult to hire, at least in Western Europe and North America (see Chap. 1). In order to further introduce the context, the main requirements for such a system are described.

For warehouses it is clear that their added cost per unit of goods should be as low as possible. This means that appropriate investment and running cost levels are required. In general, the running cost level should be lower for the automated solution in comparison to the comparable system with human pickers, but the initial investment level will probably be higher. A reasonable return-on-investment period, mostly in the order of a few years, is a measure for the marketability of the solution. Naturally, any customer will require a solution for his complete business process.

[1] Box-shaped is different from being a box: a shrink-wrapped set of six bottles of soft drink is also considered to be box-shaped.

In the case of ACP this means that whenever not all types of cases can go through the automated part, a manual part has to be added to the total solution.

The main process of a warehouse has to fit delivery frequencies and times. This means that next to the performance of the warehouse in terms of the number of orders that can be processed per day, the order completion times are relevant as well. The latter ones determine whether or not shipment time slots of customer orders will be met. For an automated system such as ACP, the system-level control should take care that the performance requirement is met in the operational system.

The next important requirement is the reliability of the warehouse as a whole. Warehouse equipment consists of large numbers of machines that deal with many different types of goods. The packaging quality of the goods is not at a guaranteed high level. One must realise that failures and errors will occur regularly in these systems. All the more it is necessary that these errors will not propagate through the system, but are confined in such a way that their effects will not impact the system availability too much. Amongst others, this means that occurring errors should be easy to correct. The layout and way of operating ACP will have a large impact on this.

6.2.2 Main Line of Development

During the main line of development, the solution space for possible ACP concepts has already been limited. The most significant choices that have been made as a starting point are given here. First, at the front-end of the goods flow the targeted warehouses typically receive goods on pallets which are stacked with one type of goods, a so-called stock-keeping unit (SKU). Second, at the back-end of the goods flow delivery of goods consists of collections of customer-ordered cases that are stacked on order-specific pallets. The stacking is done by a so-called palletiser that needs to receive the constituting cases in a sequence that is predetermined by the software that plans the stacking process. Obviously, the sequence requirement is a critical design constraint for the ACP system. The stacking process is also called mixed-case pallet stacking.

An essential choice has been made with respect to intermediate storage. While the smallest unit of handling is a case and the goods are received on pallets, it is yet decided to have intermediate storage of cases on trays. The reasoning supporting this decision is that such a tray-based storage enables an efficient case retrieval process (compared to retrieval of cases from pallets), while the replenishment process of trays is also quite efficient (compared to storage of individual cases). In order to harvest these advantages, the tray-based storage is accessible in two ways. First, the trays can be stored and retrieved from one side of the storage racks by well-known miniloads handling trays. Secondly, the individual cases can be retrieved by a so-called case picker, a miniload device that is able to retrieve a sequence of cases from different trays at different locations in the storage rack.

The choice of having tray-based storage requires some additional functions in the ACP warehouse. The trays have to be loaded with cases that come from pallets. Hence, unloading of pallets as well as loading of trays are two required functions.

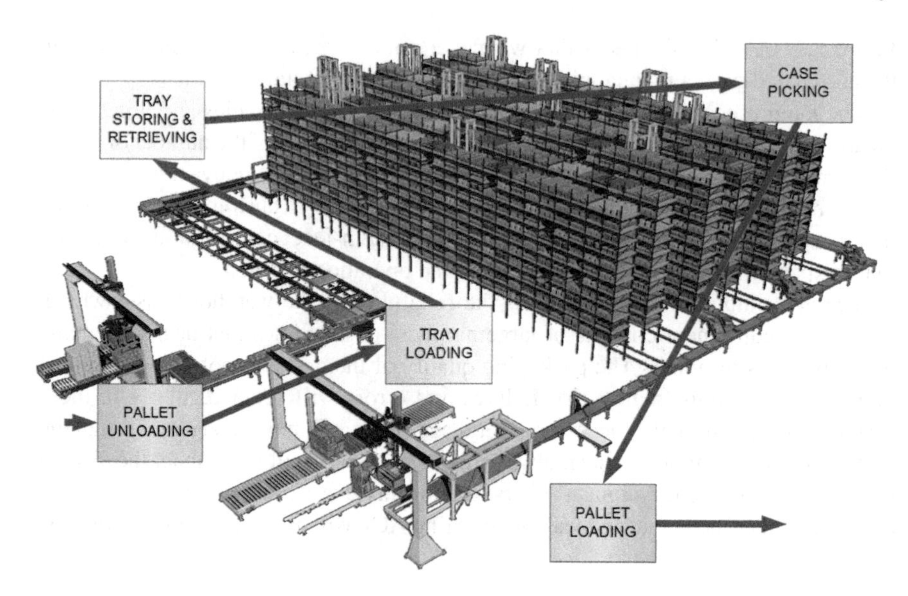

Fig. 6.1 A schematic overview of the ACP system. The most important functional blocks are superimposed on the 3D sketch. Their *colour coding* indicates which is their unit of work: *green* for pallets, *red* for cases, and *yellow* for trays. For a more lively illustration of this, see [3]

Of course, the mentioned functions cannot be realised at one place and different sorts of transport are needed to connect the functional areas: pallet transport between pallet storage and pallet unloading, tray transport between tray loading and tray storage, and case transport between case picking and mixed-case pallet stacking.

In order to be able to reason about a reference situation, a standard ACP module is introduced. This module has one mixed-case palletiser, six case pickers to match the throughput performance, and includes tray storage to serve these case pickers. A schematic overview of a module is shown in Fig. 6.1.

6.2.3 Critical Aspects of ACP

As stated before, important ACP system aspects include factors such as costs, system performance, system availability, and maintainability. Their relative weighting is not entirely clear and will probably differ from customer to customer, but also be dependent on the system concept that is chosen. At the same time, many characteristics of a number of functional components are not yet fully known: automated pallet unloading, tray loading, case picking, and mixed pallet loading are new components, firstly applied in this system.

Starting from the design context sketched in the previous section, there are tensions between the requirements. First, the required investment level of an ACP system (realisable through a minimum amount of low-cost hardware) is in conflict with the required sustained performance (which in its turn requires over-capacity of equipment and ample buffer capacity). Second, the process of building mixed-case pallets requires productive case retrieval in a predetermined sequence, which is conflicting with the necessary flexibility of retrieving cases anywhere from the storage, given the fact that one wants to distribute this retrieval over the six case pickers. Third, the investment level also confronts the availability of the system. Availability requires graceful degradation of system performance, and therefore flexible task allocation and redundancy, which is not in line with a minimum amount of hardware in order to minimise the investment level. The last tension to be mentioned here is the apparent tension between the customer's business process flexibility and the cost of the system-level controls. In Chap. 2 this issue is discussed extensively.

6.3 Examples of Models During Development

In this chapter, two examples of system aspect models that relate to the critical aspects of ACP are discussed. They relate to the second and third tension between system requirements mentioned in the previous section. Each of these tensions, at its own time, was considered the most critical issue of ACP. Therefore, these tensions were in focus of the ACP development team in different time periods, and hence it should be realised that the two illustrated models have been made in slightly different situations of the project. The first model is used to investigate the influence of sequencing on system performance. The second model allows studying system availability as a function of component availabilities.

6.3.1 The Effect of Sequencing on System Performance

One of the critical points of the ACP concept is the throughput effect of imposing a strict sequence order on the cases that arrive at the palletisers. In general, such a constraint in a system can cause severe performance drops. The main available degree of freedom at this point in time is the layout of the system, describing how the equipment is connected to each other and the dimensioning of that equipment in terms of buffer lengths and storage capacities. The introduced ACP module is defined such that the throughput capacities of the individual components are balanced.

Consider several case pickers retrieving cases for the palletisers that request their cases in sequence. The two palletisers are considered as one large palletiser, as it is assumed that all case streams are merged for efficiency reasons. The situation that arises is schematically shown in Fig. 6.2. It is to be expected that the way case pickers are connected and how much buffer space is inside the system has an influence on the throughput that can be obtained. Hence, the main question that should be answered is: how large is the adverse effect of requiring a predetermined sequence and what are the effects of layout and sizing on this?

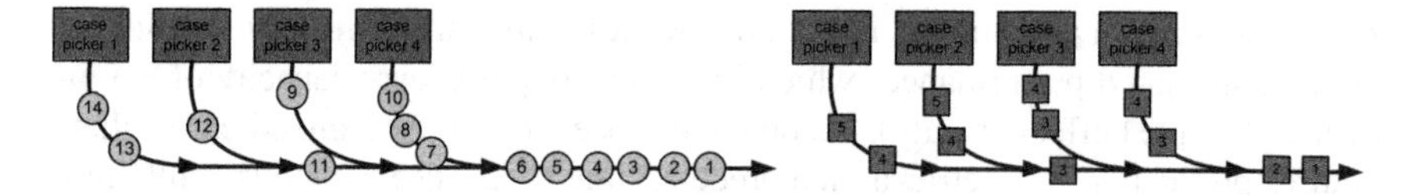

Fig. 6.2 An illustration of the sequencing requirement. Case pickers have to wait in order to keep sequence. On the left-hand side individual cases are numbered and tracked from begin to end, on the right-hand side complete pallet contributions are numbered and combined when passing merge points. The latter representation is used in the model

A model has been built to answer this question. First of all, the required kind of model is considered. If possible, a static, formula-based model would be preferred, because it contains the insights about this system aspect in a single phrase. However, such a level of insight is not readily available. Therefore, a dynamic model approach is taken. At a very high level, the system entails many independent components that synchronise with each other at their physical interfaces, i.e. they exchange cases with each other. This behaviour matches POOSL [5] very well, which is one of the many discrete event model formalisms with simulation time, and for that reason this formalism was selected. POOSL is based on a probabilistic extension of the process algebra CCS [1].

In order to keep the model as simple as possible (and thus very quickly to make) a number of strong simplifications are made. The model only takes case picking and the sequencing requirement of the system towards the palletiser into account. Travel times are left out, because they do not influence the system throughput. Even sequencing itself was simplified: mergers (merging multiple input streams of cases to one output stream) can only proceed with their work for the next pallet when all input contributions are completely available—see Fig. 6.2 for an illustration of this. The time needed to complete the contribution to a pallet is taken proportional to the number of cases involved and based on the estimated average design capacity of the case picker or merger. The relevant parameters of the model are in the degrees of freedom sketched above: the layout and dimensioning of the equipment.

Making a model should be complemented with performing analyses with it. In order to perform an analysis, the system load should be specified as well. Also here simplifications are introduced, although the input is rooted in the real world, i.e. a data file of a real prospect customer is used. The data file reflects an inventory file of SKUs and in which part of the storage area they are located, the pallet compositions, and the number of cases of each SKU contributing to the ordered pallets. In the model, it is assumed that items of a particular SKU for one pallet are picked by one single case picker. This is a worst-case estimation, because some of the SKUs will be located in multiple storage areas, i.e. be accessible to multiple case pickers. The customer data files are used to create a probability distribution to sample the sizes of these pallet contributions of each case picker. The setup for analysis around the model is shown in Fig. 6.3.

The varied parameters are taken from the set of layout options available in the ACP development team at the time of modelling. They are shown in Fig. 6.4.

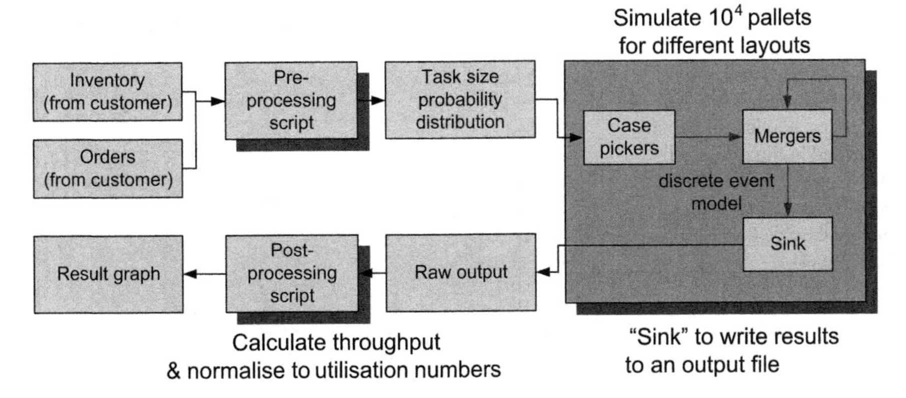

Fig. 6.3 The model context to study the sequencing effect on performance

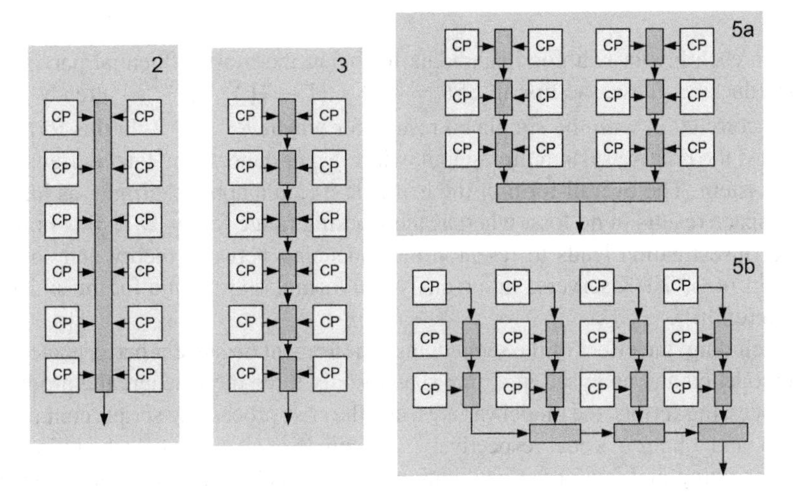

Fig. 6.4 The ACP layouts that are investigated with the sequence model. The *yellow blocks* indicate case pickers, the *orange blocks* mergers, and the *last outgoing arrow* of each layout represents the palletiser. Note that on each arrow a variable amount of buffer space is available

The main result graph is shown in Fig. 6.5, where relative throughput is shown as a function of both the buffer size (for how many different pallets the work can be temporarily stored before each merger) and the layout variant (labelled according to the labels in Fig. 6.4). The trend over different buffer sizes is clearly visible as well as the dependencies of the throughput on the layout and the interaction between the buffer sizes and the different layouts. The relative throughput level of 80% is an indicative lower bound for an acceptable system design, and hence, this graph implies that ample buffer size has to be taken into account as well as it does rule out layout alternatives.

A sensible approach to further deepen insight is trying to understand the dependencies of the results and pour that into a heuristic formula. The total amount of buffer

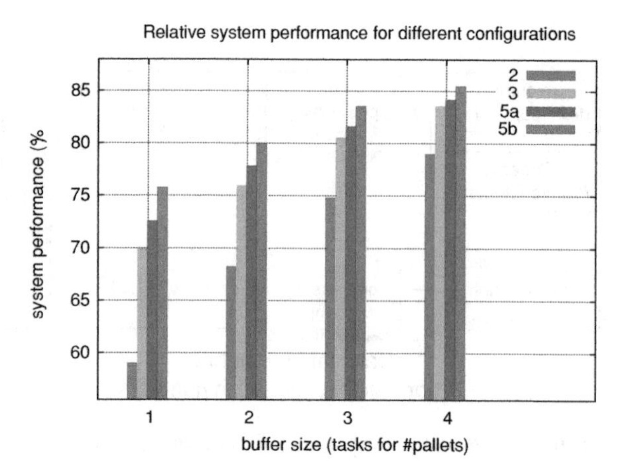

Fig. 6.5 Result graph of the sequencing model. The relative system throughput is shown as a function of buffer size and layout variant (sketched in Fig. 6.4). Full system throughput (100%) is always set equal to six times a case picker throughput

space in each layout is a good candidate for being the most influential parameter. As an educated guess, the first fit of $\theta = 100 \times \left(1 - \frac{4}{3}\{\sum_i b_i\}^{-1/2}\right)$ already gives a good correlation with the simulated results for which $R^2 = 0.9$. In this formula, θ denotes the relative system throughput, while $\sum_i b_i$ denotes the total buffer space in the system. The overall form of the heuristic formula appears alright, as infinite buffer space results in no loss, whereas the working range for $\sum_i b_i \downarrow 0$ is limited. Further investigation leads to research into queueing network theory and solving networks of G/G/1/K servers, but no analytical formulas are found for the situation at hand (cf. [6]).

Concluding, the effect of the sequencing requirement on system performance has been modelled and analysed in a time span of only three days, where the produced pre-processing scripts, the model variants, and the post-processing scripts contain 40, 95–165, and 7 lines of code, respectively. The simple analysis reveals the influences of buffer sizing and layout variants with respect to each other, i.e. it increases the insight in the system and yields guidance to what layout decisions have to be taken. It appears that still absolute figures are instrumental in steering towards final decisions: the mentioned 80% throughput level relates to a figure of merit known to experienced system designers and serves as a bottom line, irrespective of the simplifications and assumptions that have been entered into the models.

6.3.2 From Component Availability to System Availability

The second model discussed in this chapter focuses on a next critical point of the ACP concept which is the effect of component failures on the overall system availability. The reliability of case pickers is a point of attention according to the development team. There are two reasons for this: the case pickers are being newly developed and the field data of existing crane systems handling totes show non-perfect mean-time-

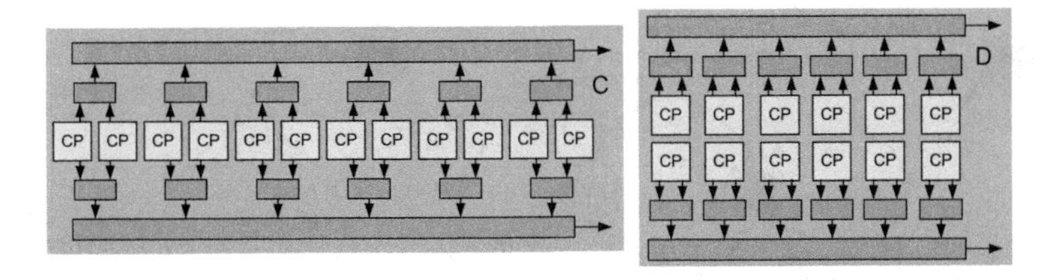

Fig. 6.6 The ACP layouts that are investigated with the availability model. Note that in layout C all case pickers deliver to both palletisers, while layout D reflects twice a half system consisting of six case pickers and one palletiser only

between-failures (MTBF [2]) figures. Due to interdependencies of the components it is not trivial to oversee the influence of case picker breakdowns on the total system availability. One has to realise the topic of this section occurred later in time than the topic of the previous section. Due to reasons not discussed here, two palletisers, each serving half the module capacity, are considered separately. The available design options have degrees of freedom in the choice of layout, but also in the choice of order planning strategy.

In the same fashion as in the previous section an as-simple-as-possible model is sought to answer the question about the influence of layout and order planning strategy on the system availability. A very simple model can be made with standard calculus of probabilities applied to handling and storage systems [4], but this does not include the possibility for investigating the influence of the order planning strategy. In order to do this the model of the previous section is extended with an order generation element and probabilistic breakdowns of the case pickers.

The layout variants that are studied are shown in Fig. 6.6. Two order generation variants are combined with these layouts. In layout C the work for a broken case picker is divided over the other case pickers, while in layout D half the system is blocked for the duration of the breakdown. The way in which case picker breakdowns are modelled is shown in Fig. 6.7. The execution of a task can be interrupted by a breakdown. The durations of breakdowns and normal operation times are sampled from MTTR (mean-time-to-repair) and MTBF distributions, the MTTR having an assumed average of 3 min and the MTBF being chosen to match the case picker availability parameter (according to $a = \text{MTBF}/(\text{MTBF} + \text{MTTR})$).

In the analyses, the same task size probability distributions as used with the previous model (cf. Fig. 6.3) are used as input. Next to this the scenario of dividing each pallet evenly over all case pickers is considered for each of the layouts. Lastly, the prediction of the probabilistic model [4] is taken into account in which the system availabilities are equal to x^1 and x^6 for cases C and D, respectively. Here x denotes the case picker availability. An overview of the results is shown in Fig. 6.8. One observes the similarity between the second and third scenarios for each layout, but also the strong deviation of first scenario from them. The preference for layout D over C is driven by this last observation and the fact that a system availability below 90% is hardly acceptable.

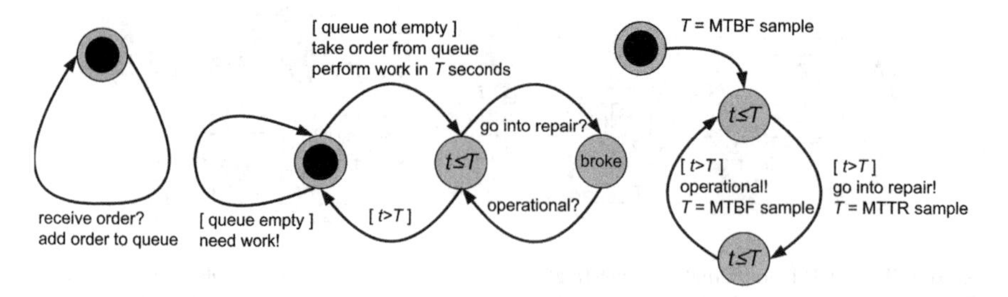

Fig. 6.7 An automata representation of modelling the breakdowns of the case pickers. The left automaton represents receiving orders. The middle automaton represents executing orders interruptible by breakdown. The right automaton represents the status of the case picker, i.e. either up-and-running or in-repair

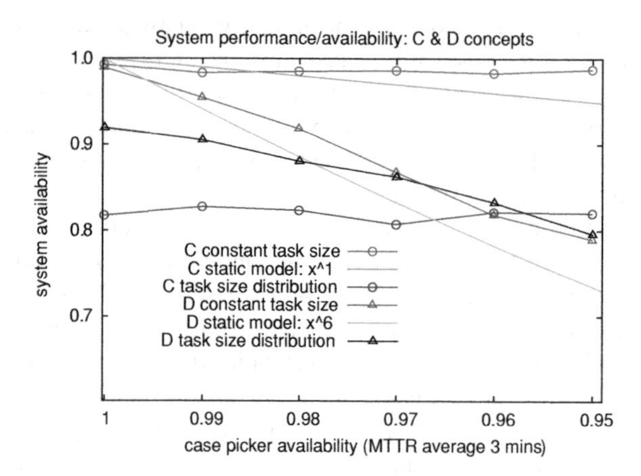

Fig. 6.8 Result graph of the availability model. The system availability is shown as a function of component availability. The different lines indicate two layout variants, each for three different scenarios. The *static model lines* indicate the probabilistic expectation of system availability expressed in case picker availability

Concluding this case, two medium-sized models (about 650 lines of code each) have been made in 2 weeks. The results show the influence of the layout variant, the variation in terms of system load, and the order planning strategy. The alignment with the static probabilistic model has strengthened the conclusions of the modelling activity: while more trust in a simpler formalism is gained, the limitations of that approach are also clearly shown. Like in the previous case, absolute figures rule the design decisions that have to be taken.

6.4 Reflection

Two examples of models supporting new concept development have been shown in this chapter. In the presented cases the system was not there yet, but by making explicit quantitative assumptions and simplifications relevant "what-if" questions could be answered by analysing the effects of varying the relevant input parameters. This approach yields relatively unbiased information about the system that helps to increase the insight in the system that is being developed.

Revisiting the first hypothesis, it is clearly shown that the approach is feasible. The effort that went into model construction and analysis was very limited, while the results were instrumental in taken design decisions and building insight. It is difficult to provide evidence on a quantitative basis, but at least the time spent fitted the period the development team was working on the critical aspect. Without models this had certainly been impossible. The value of the approach can be estimated by observing the impact on the team. Such an observation indicates that the impact of focused modelling well surpasses the common alternative, i.e. an experience-driven development process of free-format discussions. The tangibility by having explicit starting points, models, and results does play an important role in this.

The second hypothesis about relative results providing more insight than absolute figures needs some reflection as well. It has been illustrated that the variation of input parameters shows how the system works. Nevertheless, the absolute figures, whether or not they are very coarsely estimated with strong assumptions, connect to the experience of the developers. One should be very well aware that this influence can have an adverse side-effect to the intended message itself: another message might be implicitly communicated as well while displaying results.

As a last reflection, it is observed that the simplifications made during modelling have to match the level of insight shared by the development team in order for the results to be believed. In many instances this was the case: the order planning by a stochastic worst-case representation, the simplification of the sequencing requirement into a start condition, the omission of transport times, and the stochastic representation of breakdowns of case pickers, are all examples where the simplification was explainable and accepted. Buffer sizes expressed in amount of work for a number of palletisers was hardly accepted. The probabilistic model for system availability was also not within the frame of common reference: the results of this model actually were available to the team before the shown simulation model, but not commonly accepted at that time.

To conclude, the judgment of the level of criticality of design issues is independent from the modelling itself. It requires a flexible mind of system developers to model what is needed to answer a question instead of just modelling what can be modelled. Sometimes, it requires a paradigm shift in the environment this is done. In this chapter, it is demonstrated that this can be applied in an economic way, properly balancing the costs and benefits of modelling, focus sing on building insights of the system's nature, and staying sufficiently close to the body of experience that is already present in the development team.

References

1. Bergstra JA, Klop JW (1984) Process algebra for synchronous communication. Inf Control 60:109–137
2. Johnson BW (1989) Design and analysis of fault-tolerant digital systems. Addison-Wesley, Boston

3. Vanderlande Industries (2011) Automated case picking. http://www.vanderlande.com/Distribution/Products-and-Solutions/Order-Picking/Case-picking/Automated-Case-Picking.htm, Viewed April 2011
4. VDI-Fachbereich Technische Logistik (1992) Anwendung der Verfügbarkeitsrechnung für Förder- und Lagersysteme. Richtlinie VDI 3649, VDI-Gesellschaft Produktion und Logistik, Düsseldorf
5. Voeten JPM, van der Putten PHA, Geilen MCW, Theelen BD (2007) SHE/POOSL website. http://www.es.ele.tue.nl/poosl, Viewed April 2011
6. Whitt W (2004) Heavy-traffic limits for loss proportions in single-server queues. Queueing Syst 46:507–536

Chapter 7
Warehouse System Configuration Support through Models

Roelof Hamberg and Jacques Verriet

Abstract The warehouse system sales process leads from a specific customer request to a specific customer quotation. This process of configuring a warehouse system with existing components involves a sequence of steps which contain increasingly more details. In this chapter, it is shown by two examples that easy-to-use essential system simulation models can be applied early in the sales process with a good cost/benefit ratio. Configuration of the models and their analyses require little effort, while their results are comparable to detailed simulations. Next to these examples of specific models, an overall sales process improvement is proposed by consolidating the available knowledge in models through a framework. The basic concept for this evolvable framework is to provide traceability of customer data and results throughout the sales process.

7.1 Introduction

In the market for warehouses, a significant part of the system design process takes place before actually selling the system. In the first phase of engineering, called sales engineering, a customer-specific solution, i.e. a customised warehouse design, is proposed. A sales agreement is based on such a design, which for the customer should match the requirements and for the supplier should be economically feasible to realise. The large variation of systems in the market, all being unique systems, each fitting specific customer business processes, poses the sales engineers with a

R. Hamberg (✉) · J. Verriet
Embedded System Institute, P.O. Box 513, 5600 MB, Eindhoven, The Netherlands
e-mail: roelof.hamberg@esi.nl

J. Verriet
e-mail: jacques.verriet@esi.nl

R. Hamberg and J. Verriet (eds.), *Automation in Warehouse Development*,
DOI: 10.1007/978-0-85729-968-0_7,

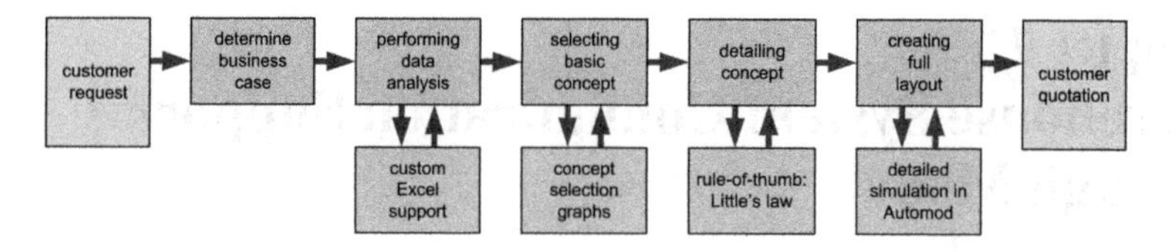

Fig. 7.1 An overview of the general steps in the sales process, going from customer request to customer quotation. At the *bottom* row some of the support methods are indicated

challenge: how to optimally configure a realistically feasible system within limited time and resources?

In the current situation in Vanderlande, a number of methods is available that assist the sales engineer in configuring a system from a large set of known components. Examples of such methods are the analysis of customer data to obtain their main characteristics, graphs that relate customer order characteristics to primary warehouse concepts, estimation of the effects of warehouse design choices on system performance figures through rules-of-thumb, and detailed simulations which can be built on the basis of proposed warehouse layouts and their rules of operation. In Fig. 7.1, an overview is given of the typical steps that are taken in the sales process, along with the mentioned support methods.

An often recurring problem with the currently available methods is their appropriateness. The simple methods that can be applied in very early phases, are actually too simple and neglect the influence of relevant factors. This has been regularly indicated by Vanderlande system engineers. On the other hand, detailed simulations are very costly to make and can only be made late in the process, when the process already has converged to a single design variant. This has also been recognised by Vanderlande. Finally, having an up-to-date view of all available possibilities is a challenge by itself. Several initiatives within Vanderlande have addressed the indicated problems, but their results are not widely known nor accessible amongst the engineers.

In this chapter, the goal is to show that the application of so-called easy-to-use essential models can help to provide a solution to the aforementioned problems. These models should be applicable in early phases, because this is most efficient. The efficiency in itself is important, because the sales phase is a pre-investment phase of the warehouse system supplier that should not take too much time nor effort. The need for applicability in early phases is based on the fact that early design choices have a larger impact on the complete design and are less costly to change than late design choices.

The hypotheses that underlie this work are the following. First, it is supposed to be feasible to create relevant easy-to-use essential system models that support early validation by sales engineers. Such models will have to be based on a simulation approach, as analytical methods for dealing with the required degree of variation that is observed in warehouse systems, are not known. Second, the integration of scattered pieces of knowledge into one framework is supposed to increase the quality of the sales process. Such a framework should be able to cope with future changes in required systems, available components, as well as a variety of supportive models.

7.2 Black-Box System Performance Models

The main idea of easy-to-use essential system models is that relevant system aspects are modelled in such a way that analyses can be done with limited effort. In general, the effort consists of the time spent to make the model and the time spent to perform the analyses. In the case of black-box models, the model-creation time is needed only once: the assumption of this approach is that a class of possible systems can be covered in one generic model. This only pays off if the required analysis time, including configuring the model, is shorter. Of course, the construction of the generic model itself should also be within reasonable limits.

In the configuration of a warehouse system the relevant system aspects are manifold. Some examples are initial system cost, running system cost, system throughput, customer order completion times, system surface and volume, and system availability. For a number of these system aspects, the relation between their value and the system composition is quite trivial: for instance, the initial system cost is just a sum of all the constituent components. Other aspects are the result of non-trivial interactions across system components. A good example of this is the system throughput, which not only relates to the throughput capabilities of the components, but also on order patterns, buffer sizes, and control strategies. It is this system aspect, throughput, and the closely-related properties order flow times, work-in-process levels, and component utilisation levels, that will be in focus in this section.

An approach to provide the system performance analysis for throughput-related aspects in an easy way is to construct a multi-parameter black-box model that allows experiments to be conducted in a flexible way. This is illustrated in Fig. 7.2, where the left-hand side denotes the environment for user input, the right-hand side denotes the environment for result output, and the generic model is in the middle. In the context of the Falcon project, this approach was employed several times. Two different applications of that approach will be outlined in this section.

7.2.1 CPS Simulator-in-a-Box

The first system performance model originates from a request to enhance the estimation of throughput performance of compact picking systems (CPS) in an early phase of the sales process. An example of CPS is shown in Fig. 7.3. To estimate the throughput as a function of the chosen system configuration, simple rules-of-thumb (such as Little's law [4], that relates average throughput, average flow time, and average work-in-process level) yield inadequate information. This situation has worsened with the introduction of more advanced scheduling of jobs, which not only affects accuracy of the existing rule-of-thumb approach but as a result also its credibility.

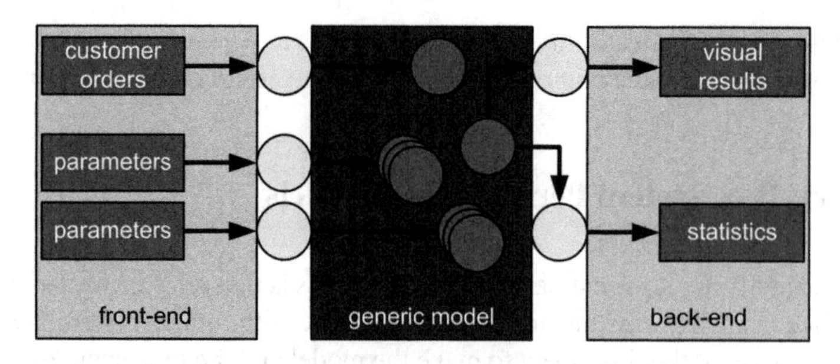

Fig. 7.2 Black-box view of a model with the input parameters and the output results. The *circles* denote internal parts of models and interfacing artefacts, such as intermediate files

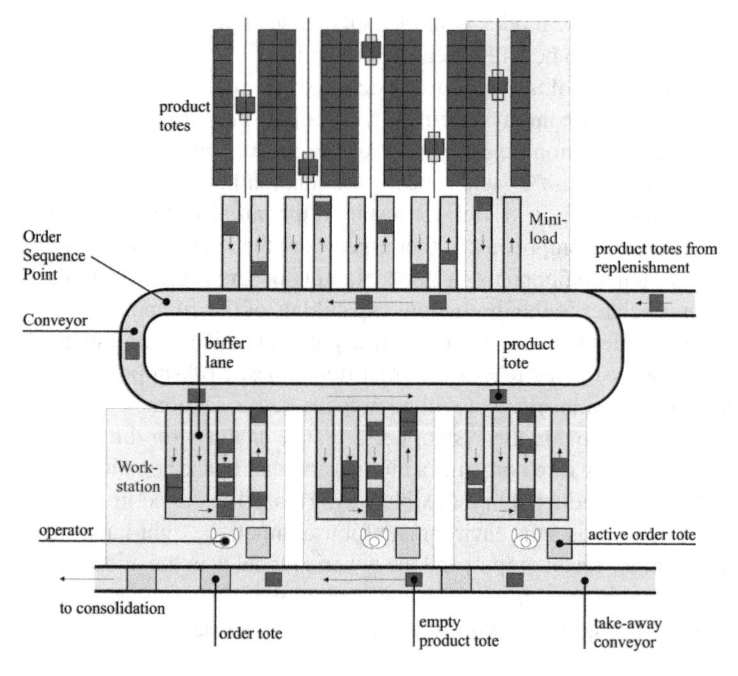

Fig. 7.3 An example of a compact picking system (CPS). An area for automated storage and retrieval system (ASRS) with product totes (*at the top*) is coupled via a transport loop to a number of manual picking workstations (*at the bottom*). Completed order totes are taken towards the consolidation area via a separate conveyor

The challenge of enhanced modelling support for this case has been addressed by building a generic CPS simulator. This simulator can be controlled by a Microsoft Excel worksheet, which also retrieves the results of a simulation run, i.e. result statistics and optional visualisation.

One of the central challenges of the design of the simulator is to find suitable system and component representations that serve the modelling goals. The main choices are:

- Customer orders consist of order lines, each of which can be completed ultimately by picking items from one product tote into an order tote. A product tote has to be retrieved from an ASRS, to be transported over the loop, and to be picked in the workstation. These three activities are called tasks.
- For throughput-related aspects, the time period Δt_i to fulfil a task i forms a simple and sufficient basis for the model. Layout details are less relevant, although they obviously influence the time periods.
- The exact task details, such as involved product types and product locations, are ignored. Their influence is modelled as variability on the task completion times, which is considered to be sufficient. For every task, a batch of product totes is handled in a time period Δt that only depends on the batch size N. The variability in Δt due to task details is captured by defining it as a sample from a probability distribution: $\Delta t = \text{sample} P_{\Delta t}(N)$.
- The absence of product types and locations implies trivial order planning: only the number of product totes within each order is required in the model. These numbers are taken from a user-defined distribution, based on customer data.
- The sequence in which tasks are performed as well as the waiting times that result from limited buffer space, are relevant for system throughput. These properties of the system are hardly abstracted in the model, i.e. the load balancing logic in the existing warehouse control is taken into account quite literally.
- The replenishment and return streams of product totes are not modelled explicitly. The return totes are assumed to be included in the task times mentioned before. The effect of replenishment is negligible because the number of items per product tote is large.

Together, these considerations result in a model that consists of a material flow layer (that is built from a series of queues and servers), a load balancing layer (that reflects the negotiation protocol to determine the sequence of tasks), a simple order generator (to provide customer orders to the model), and a simulation controller (to log simulation data and statistical information). Such a model has been implemented in POOSL [10], and has a structure as shown in Fig. 7.4.

The CPS simulator is equipped with a Microsoft Excel worksheet in order to be able to control it in a simple way. The control consists of providing input parameters and starting the simulation with these parameters. The set of input parameters is shown in Fig. 7.5a. The terminology is kept close to the vocabulary of the sales engineers. The majority of parameters concern the physical components. Loop buffering, number of assigned orders, and order size distribution are relevant parameters for the order planning and load balancing (scheduling) parts of the model.

After a simulation run, the results are automatically collected in a separate worksheet. The available results cover throughput and utilisation levels, work-in-process levels, and flow times for all components of the system. These results are collected by the model itself through cumulative observations during simulation. Each obser-

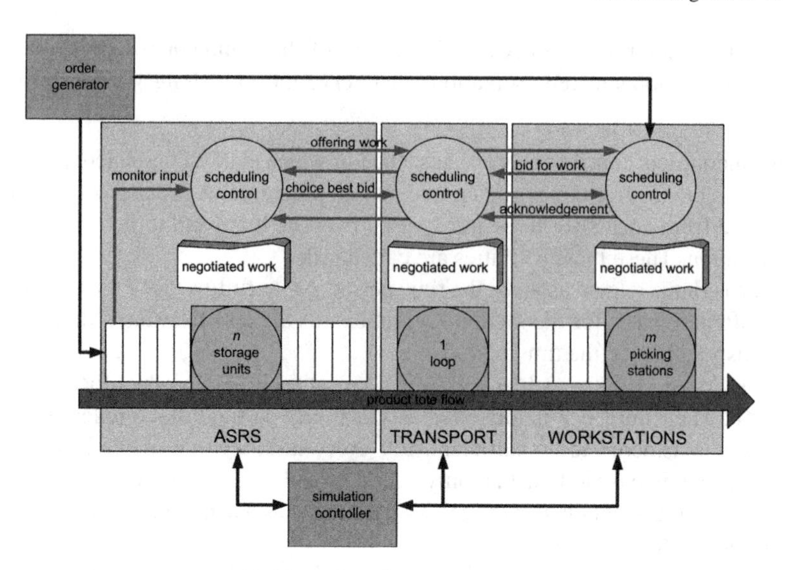

Fig. 7.4 Schematic model structure of the CPS performance model. The ASRS, transport, and workstation areas have identical structures consisting of a material flow layer and a load balancing (scheduling) layer

vation is constructed from average values over a fixed time period of 10 simulated minutes. Whenever the statistics of the cumulative series have sufficiently converged, the simulation self-terminates. Furthermore, a Gantt chart of the performed tasks in the system can be visualised which reveals the dynamic behaviour of the system in terms of the coordination and timings of the tasks executed by different components. Screenshots of the results are shown in Fig. 7.5b, c.

The model has been validated in the following ways. First, the results of realistic as well as extreme test cases have been studied with a system engineer of Vanderlande, who is knowledgeable in CPS solutions. In order to verify these results, the statistics as well as the Gantt chart representations are inspected: these are found to meet the expectations. Second, the results of a detailed simulation of a system was compared to the results of the CPS simulator. This resulted in 8% difference in throughput, which is considered acceptable given the differences in the system control rules that were present in this case.

The CPS simulator has become available through IT systems at Vanderlande, i.e. sales engineers have instant access to the tool. Several real customer cases have been studied with the simulator. From the start, such studies consist of experiments with many different parameter configurations, which is natural considering the limited effort that is needed to do this. Typically, the setup of an experiment as well as a single simulation run takes only a few minutes. Altogether, this is a good illustration that easy-to-use essential system models are conceivable to provide more insight without including all details.

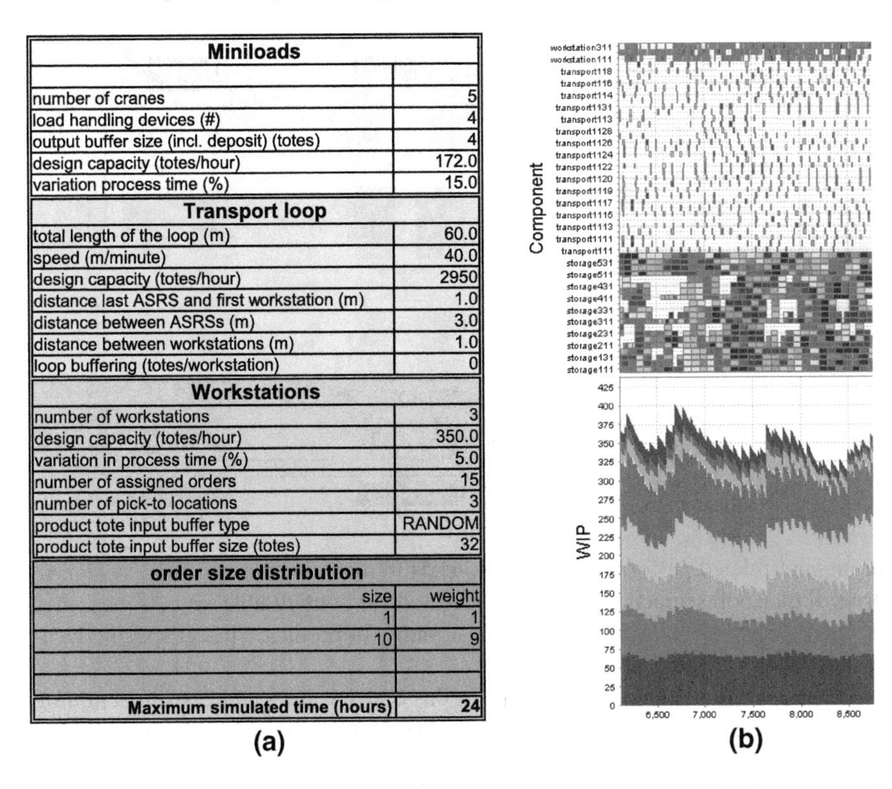

Miniloads	
number of cranes	5
load handling devices (#)	4
output buffer size (incl. deposit) (totes)	4
design capacity (totes/hour)	172.0
variation process time (%)	15.0
Transport loop	
total length of the loop (m)	60.0
speed (m/minute)	40.0
design capacity (totes/hour)	2950
distance last ASRS and first workstation (m)	1.0
distance between ASRSs (m)	3.0
distance between workstations (m)	1.0
loop buffering (totes/workstation)	0
Workstations	
number of workstations	3
design capacity (totes/hour)	350.0
variation in process time (%)	5.0
number of assigned orders	15
number of pick-to locations	3
product tote input buffer type	RANDOM
product tote input buffer size (totes)	32
order size distribution	
size	weight
1	1
10	9
Maximum simulated time (hours)	24


```
                                      A
1   component  para average  [  confidence ]  accur [    min,    max_
2  ----------------------------------------------------------------
3    storage0  Thru    825.6  [ 813.4, 837.7]   1.5% ------------->     96% (AVERAGE over storages_
4              flow   1497.2  [1447.2,1547.2]   3.5%
5              WIP     352.9  [ 319.3, 386.5]  10.5%
6    storage1  Thru    162.1  [ 156.3, 167.9]   3.7% [   68.1, 256.5]------------->     94_
7              flow   1442.4  [1379.1,1505.6]   4.6% [ 343.7,2159.3_
8              WIP      66.7  [  59.5,  73.9]  12.2% [   35.0,   98.0_
```

(c)

Fig. 7.5 A sketch of the input parameters of the CPS performance model **(a)**, the optional Gantt chart visualisation **(b)**, and a part of the output statistics **(c)**. The layout of the input parameters is adapted for convenient representation here.

7.2.2 The Generic Transport Routing Simulator

The second example of a system performance model originates from the need to investigate the effect of system dimensioning parameters and order tote routing on the performance of a system with roaming shuttles (RPS). In this system, the customer orders are picked in the storage area, and order totes with the picked items are moving from one storage location to another. Such a system closely resembles zone picking systems (ZPS), in which order totes visit several workstations with local storage in a sequence. A sketch of the ZPS concept is shown in Fig. 7.6. The observation that routing of order totes is a main factor in determining the system

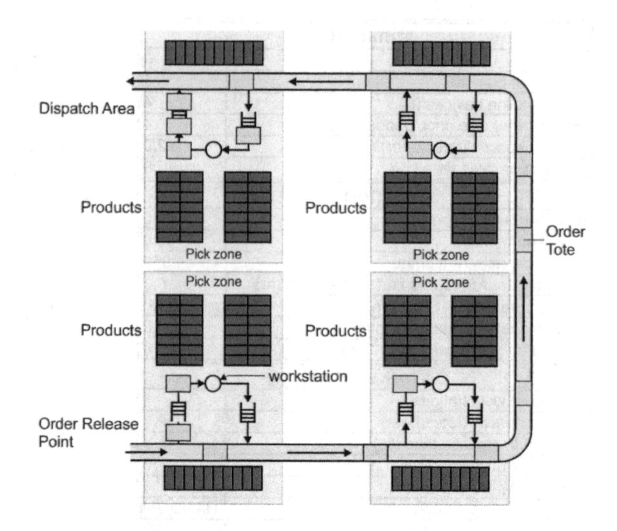

Fig. 7.6 An example of a ZPS configuration. The order totes visit different workstations, each of which deals with a specific set of products

throughput-related performance aspects, leads to a representation of the system in terms of transport segments which are connected in a configurable way. As the routes have to be constructed from the tasks, positioning of product types at specific locations is relevant for this approach.

The challenge of modelling support for this case has been addressed by building a generic transport routing simulator. This simulator can be controlled by a configuration file, while output statistics are collected in a file and the simulation itself is visualised in 3D. The latter visualisation can be stored as well for later inspection.

The main choices for construction of the model that serves the modelling goals, are discussed here:

- Customer orders are fulfilled by a set of order totes. Each of these have to travel along a series of locations to collect items from stored product totes. Each series of locations is called an order tote route.
- The modelled system is built from transport segments, defined in 3D by begin and end points, and order totes, which travel over the transport segments along their routes.
- Time progression in the model is represented by iterating over system states, comprising of active customer orders, route allocations, and the locations of all order totes. Each new iteration represents a fixed time step, which is set to 1 s here. This yields sufficient accuracy.
- Buffers are represented by segments. The buffer sizes are captured by a logical length, independent from the distance between begin and end points. This length reflects the buffer size.
- The actual throughput of each segment is captured by an activity frequency, which serves as a correction on the default speed of 1 logical length unit/logical time unit.
- For activities other than transport, an activity time is introduced. It adds additional process time to a segment, representing tasks such as picking, storing, and receiving.

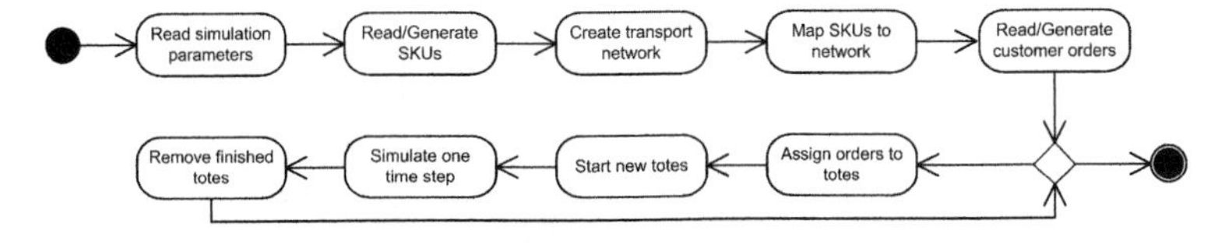

Fig. 7.7 State machine of the generic simulator

- Segments can have multiplicities, which leads to multiple independent, parallel segments that have identical properties.

These considerations result in a model that consists of a transport network (containing the physical infrastructure of the system), a collection of order totes, an order generator (to transform customer orders to transport routes), and a simulation controller (to log simulation data and statistical information). Figure 7.7 shows the state machine of the generic simulator as it was implemented in Java. The state machine has an initialisation phase in which the SKUs, layout, and orders are created from a file or a probability distribution. After the initialisation, the simulator performs a sequence of actions for each time step. This involves the creation, simulation, and removal of order totes until all orders have been fulfilled.

The usage of the model has been facilitated through a few specific measures. First, the parametrisation of behaviours is specific for this model. It is made easy through the possible specialisation of abstract classes. This parametrisation is available next to the standard parameters such as number of components, sizes, etc. In general, these behaviours reflect heuristic rules for control strategies, such as order release sequence, choice of pick locations, sequence of SKU planning, and a preferred sequence of locations to visit. These abstract classes are similar to the skeleton behaviours used in Chap. 2. Second, for each class of systems, such as RPS and ZPS, a template approach facilitates the generation of the transport network: the layout does not have to be drawn by hand, but rather is generated on the basis of some key parameters. A custom network can also be made, which provides more flexibility at the cost of spending more effort on model specification.

Analysis of model under test can be performed by executing the model with a specific configuration file with input parameters. The results can be studied numerically, and visualised in a custom, lean 3D visualisation program that can be connected to the Java model. A screenshot of the visualisation, which was developed using OpenSceneGraph [6], is shown in Fig. 7.8. The visualisation can be enabled during a simulation run, but also a replay of logged simulation runs is possible. Summary statistics are optionally represented as different colours on the generated system layout.

The model has been generalised to cover CPS solutions as well by interchanging the roles of order totes and product totes [9]. This CPS variant of the generic transport routing simulator was validated with an extensive simulation model based on AutoMod [1], including a detailed layout of an existing retail warehouse. All relevant performance statistics (system throughput, utilisation levels, buffer occupancy

Fig. 7.8 A 3D visualisation of a large ZPS setup. *Green* totes are moving, *magenta* totes are waiting, *yellow* totes are active in picking. The occupation level of the buffers as well as transport behaviours can be observed instantly

levels) were within a few percent of each other, which is within the standard deviation range of the estimates themselves.

The usage of the model indicates that comparable insights about system performance can be obtained with much less effort than needed for detailed simulations. Typical configuration setups take only minutes, individual simulation runs even less than one minute. This facilitates more extensive exploration of system configuration possibilities leading to improved insight in parameter dependencies and better solution propositions.

7.2.3 Reflection

Two examples of performance models have been shown in the previous sections. Depending on the system characteristics at hand, different simplifications have been made, although both of them take task fulfilment time as leading characteristic of the constituting components. Both models can be applied in an early phase of system configuration, typically when the sales engineers are constructing the propositions for their customers. As opposed to the current way-of-working, a detailed layout is not needed. The main advantage of this is that multiple configurations can be studied in a fast and easy way, earlier in the process, which improves the quality of the design decisions of the sales engineers.

There is a trade-off between the flexibility of the models and their configuration times. As long as the system under study fits the generic underlying model (with provided templates), configuring the model as well as performing simulation runs is very fast in both cases: it is a matter of minutes per case. If one wants more flexibility, such as building custom networks or implementing specific behaviour rules, the setup of the model generally takes several hours to days. This is possible with the second model, the generic transport routing simulator.

Systems with non-standard components and/or non-standard control algorithms are not only more expensive to build, they are also more expensive to study when they do not fit the model simplifications. This directly relates to a sound system architecture: the degrees of flexibility of a standard product family should reflect the needs in the market. Other system variations are much more expensive and therefore not recommendable. Each of the two examples illustrates a different choice as to what flexibility is still available in the model, but both show that it pays off to have these degrees of flexibility clear.

7.3 Integrated Warehouse System Configuration Support

Black-box system models provide a good opportunity to support warehouse system configuration for a single, dedicated purpose. However, in order to fully support warehouse system configuration, assistance should be provided to each of the consecutive steps in a design process that are taken to come from customer contact to a full system quotation, and eventually a full implementation of the system. The steps of the first part, the sales engineering process, have already been globally shown in Fig. 7.1, and proceed from coarse grained to very detailed. For all of these steps, it appears that scattered knowledge and models, related to existing components and system concepts, are available in the context of Vanderlande. Sales engineers are the main players managing this process from begin to end, and they have a challenge in keeping an up-to-date overview of the support that is available for the full spectrum of activities they have to perform.

The approach here is to acknowledge the presence of many different tools and the fact that new ones will be made as well. Hence, it is not tried to transform them to one single tool, but rather to integrate them at the level of data exchange and synchronisation. An integration framework supporting the process in this way, should also adhere to the prevailing sales process as shown in Fig. 7.1 in order to seamlessly fit the workflow of the sales engineers. A prototype of such a framework is presented in this section.

7.3.1 The Warehouse Design Toolbox

The Warehouse Design Toolbox (WDT) prototype [3] provides an umbrella over four of the five process steps shown in Fig. 7.1: it starts with the (customer) data analysis step, as the financial aspect is not yet taken into account in this prototype. In WDT's user interface, shown in Fig. 7.9, the different tabs support consecutive steps of the sales process. An overview of the tabs' contents is given here:

Customer Data The customer data files, consisting of an overview of all products and typical order files, can be imported to WDT's database in a flexible way.

Data Filtering and Analysis Analyses on the customer data can be performed, e.g. filter classes of products, select orders from a certain period, get a graphical overview

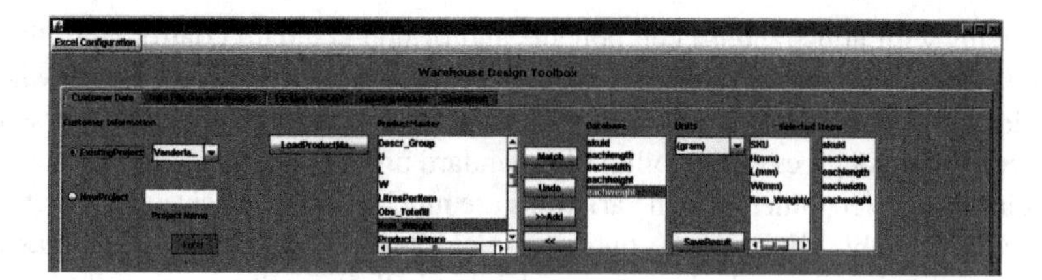

Fig. 7.9 A snapshot of the user interface of WDT

of relationships between different product and order aspects. Filters can be stored for later reuse.

Normally, such analysis is done in a spreadsheet application such as Microsoft Excel, or for larger data-sets in a database application such as Microsoft Access. Here, the focus is on generic accessibility of limited but essential data analysis in the context of warehousing.

Picking Concept A model to support the choice of a picking concept can be applied from here. The so-called basic concept decision-making tool was made by Van Haaster [8]. It allows the user to rank decision factors relative to each other and compiles an advice for the basic system pattern to use to build the solution. The advice includes a sensitivity indication.

Queueing Models For a number of concepts queueing models with specific analysis techniques are available. Queueing models typically do not take detailed customer data into account, but their analyses reveal whether or not proposed system configurations fulfil throughput requirements.

A queueing model for multi-segment ZPS systems has been made by Bakker [2]. This model applies an iterative method in which for each step a mean-value analysis is applied to arrive fast at overall results at the cost of a decreased accuracy.

A queueing model for CPS systems has been made by Liu and Adan in the context of the Falcon project [5]. A roll-unroll technique is applied to find approximate solutions to such networks. Also this analysis is fast at the cost of decreased accuracy.

Simulation The generic transport routing simulator discussed in Sect. 7.2 is available from this tab. This simulator takes the detailed customer data, or optionally a filtered subset of this, as its input in order to be able to analyse the system performance aspects under customer-specific load conditions.

The data filtering and analysis is perceived as an essential part. It provides a simple means to create relevant customer scenarios of (parts of) the system. The streamlining of the required functionality in this part has a positive effect on the subsequent process steps, because errors are less likely to occur.

Special attention has been given to extendability of this framework, because the supporting models and analyses will evolve over time. The approach taken is that available models are accessible through invocation of their respective tools, while their input is prepared first by WDT, and their output is processed and shown in WDT after the tool has finished. The generic design of WDT is sketched in Fig. 7.10. Central to its concept is the common database that is used to keep track of

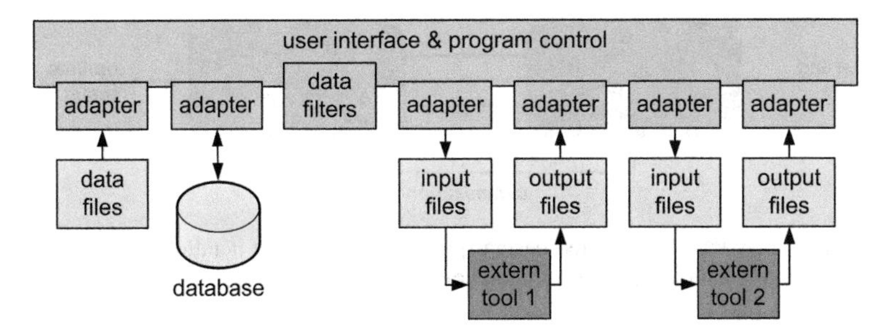

Fig. 7.10 The WDT design. Parts in *green* belong to WDT itself, the *yellow* parts are connected quite closely, while the *red* parts are external to WDT. Multiple tools are connected in this way

information within and across projects, including customer data, filter definitions, and past analysis results. Extension of WDT requires creation and integration of new adaptors, which can be done with relatively small effort. The modularity of the UI and the simple interfacing to tools via intermediate files are the main enablers for this.

While a full validation of the WDT application has not been done, the first feedback on WDT has been quite positive despite the fact it is only available as a prototype. Especially the possibility to integrate new tools and being able to gradually create a common base of knowledge are perceived as valuable opportunities. The strong points of WDT include the possibility for distributed use and the traceability of customer data through the different steps of the sales engineering process.

Future work on the warehouse design toolbox would have to include the link to the last step in the sales process, i.e. creating the detailed layout. The access to the database should be extended to support the multi-user aspect, a key issue in transforming the WDT prototype into a organisational support tool.

7.4 Conclusion and Outlook

In this chapter, three examples of supporting warehouse configuration by models have been discussed. In warehousing, such system configuration for a specific customer is done before actually selling the system, and therefore, a very efficient way of working is needed. The example models address this: the first two models illustrate how quick and accessible performance analysis can be done, while the last example illustrates how multiple models can be brought together in a comprehensive framework.

Reflection on the initial hypotheses shows that model-based support for the configuration of warehouse systems is feasible and quite effective. First, the feasibility of easy-to-use essential system models that support early validation by sales engineers is shown by two examples in the area of system throughput performance. A simulation approach, based on strong simplifications and prepared for building a system according to templates, was followed. This resulted in models which are easy to

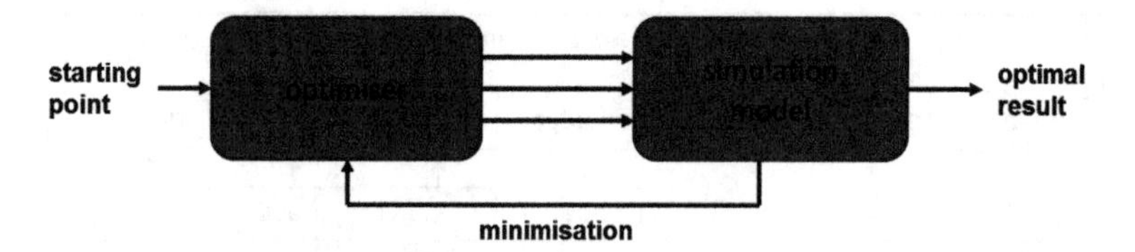

Fig. 7.11 The approach of automated design-space exploration. An iteration over many model analyses is done to obtain an optimal configuration

configure and whose simulation runs are fast. Their end results match the detailed simulations within acceptable margins.

Second, the integration of knowledge into one framework was exemplified with the warehouse design toolbox. The WDT was designed for coping with future changes, but its implementation did not exceed the prototype level. The claimed hypothesis of increasing the quality of the sales process cannot be fully validated, but the initial tests indicate a positive tendency.

The black-box models can be considered as a part of the toolbox to engineer a family of products, i.e. a product line. They belong to the sales engineer's standard equipment to be able to configure a system that fits the needs of a customer. An essential issue for future work is the consideration of the relevant level of standardisation of elements. The presented approach supports configuration of single concepts such as CPS, while an increased level of system flexibility would also require the configuration of hybrid concepts. The latter situation asks for models which are configured themselves by a system specification editor which is much more in line with the domain-specific approach as sketched in Chaps. 2 and 4. This would impact the WDT concept as well, because the specification editor would have a major interaction with this framework.

Another future extension of this work is facilitated by the inherent nature of the models that were employed. In this chapter, system throughput performance has been modelled as a black box, covering all relevant parameters and assuming smooth behaviours of the system aspects with respect to parameter variations. These models reflect a huge set of system configurations, while every single configuration can be evaluated relatively fast. With such models, the spanned design space can also be systematically searched in an automated fashion instead of by a human (see Fig. 7.11). An initiative of this kind has been addressed in the Falcon project by Reehuis and Bäck [7], using the generic transport routing simulator as a basis. Future investigations are needed in order to harvest the full potential of this approach.

References

1. Applied Materials Inc (2011) Automod website. http://www.automod.com, Viewed April 2011

2. Bakker RF (2007) Performance analysis of zone picking systems. Master's thesis, Eindhoven University of Technology, Department of Mathematics and Computer Science, Eindhoven
3. Hakobyan L (2009) Warehouse design toolbox. SAI technical report Eindhoven University of Technology, Eindhoven
4. Little JDC (1961) A proof for the queuing formula: $L = \lambda W$. Oper Res 9:383–387
5. Liu L, Adan IJBF (2011) Queueing network analysis of compact picking systems. Working paper
6. OpenSceneGraph: OpenSceneGraph website (2007). http://www.openscenegraph.org, Viewed May 2011
7. Reehuis E, Bäck T (2010) Mixed-integer evolution strategy using multiobjective selection applied to warehouse design optimization. In: GECCO '10 Proceedings of the 12th annual conference on genetic and evolutionary computation, pp. 1187–1194
8. van Haaster RAJG (2008) Basic concept decision-making for material handling systems. Master's thesis, University of Twente, Faculty of Engineering Design, Production and Management, Enschede
9. Verriet J (2011) Falcon Compact Picking System (CPS) simulation. http://www.youtube.com/watch?v=DcJs3ALbDDg, Viewed May 2011
10. Voeten JPM, van der Putten PHA, Geilen MCW, Theelen BD (2007) SHE/POOSL website. http://www.es.ele.tue.nl/poosl, Viewed April 2011

Part IV
Automated Item Handling

Chapter 8
An Industrial Solution to Automated Item Picking

Toine Ketelaars and Evert van de Plassche

Abstract Item picking is one of the most challenging functions in warehousing when it comes to automation. This chapter describes a solution for this challenge that serves as a benchmark for further research in enabling technologies. It appears that with the off-the-shelf technologies, SIFT-based item recognition, suction gripping, gantry robots, and service-based process control, a workstation can be constructed that covers almost two thirds of the required pick actions. The main class of items that cannot be picked consists of items without a flat surface or without sufficient visual features. These omissions are the starting points for alternative solutions researched in Chaps. 9–12.

8.1 Introduction

At the centre stage of many warehouse operations is the so-called order-picking process. In highly automated warehouses often a *goods-to-man* solution is used in the order picking of small items (*item picking*). The term *item* is generally used for products that are too small to be transported on conveyors individually. Items are therefore packed in transport containers, referred to as *product totes*, illustrated in Fig. 8.1a.

In a goods-to-man solution these product containers are transported to a centralised picking area. Here a human operator performs the task of a so-called *item picker* and takes out a specified number of items and places these items in an order container, called *order tote*, as illustrated in Fig. 8.1b. This central area where an item picker works is called a *workstation*.

T. Ketelaars (✉) · E. van de Plassche
Vanderlande Industries B.V, Vanderlandelaan 2,
5466 RB Veghel, The Netherlands
e-mail: toine.ketelaars@vanderlande.com

E. van de Plassche
e-mail: evert.van.de.plassche@vanderlande.com

R. Hamberg and J. Verriet (eds.), *Automation in Warehouse Development*,
DOI: 10.1007/978-0-85729-968-0_8, © Springer-Verlag London Limited 2012

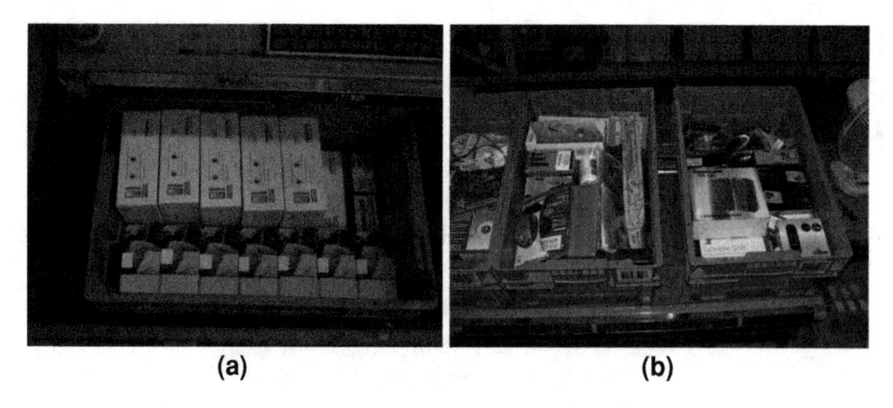

Fig. 8.1 Examples of a typical product tote and order tote. **a** Product tote with multiple identical products. **b** Order tote with different products for one order

Once all items belonging to a specific order are placed in the order container, the order container is transported further in the warehouse towards a consolidation and shipping area for further processing.

Many types of workstations exist, depending on specific criteria such as number of order lines per day, number of order lines per order, number of items per order line, dimensions of the items, etc. An example workstation is shown in Fig. 8.2.

Today, almost all item picking in retail warehouses is performed as a human task. In this chapter, it is addressed what it would take to automate item picking. Subsequently, the challenges when automating item picking will be described in Sect. 8.2, and an industrial solution based on commercially available components and their integration is proposed in Sect. 8.3. The working range of the solution will be studied in Sect. 8.4 by analysing what the proposed workstation can and cannot process, and suggestions to further improve the current "state of practice" in industry will be given in Sect. 8.5.

8.2 Automation of Item Picking: The Challenge

Automation of item picking is becoming more important as the cost of labour increases in the developed countries. As item picking is often done during night shifts and is considered hard, uninspiring work, high labour turnover poses additional problems with respect to training and quality assurance.

High levels of automation can already be found in other parts of the warehouse (automated pallet handling, high bay storage and retrieval systems). However, the demands on item picking are such that no economical solution has yet achieved a sustained market presence.

A human item picker performs a multitude of tasks which are all challenging to automate per se, let alone in combination:

Fig. 8.2 Commercial item-picking workstation (PICK@EASE.4 from Vanderlande Industries B.V., see http://www.vanderlande.com/)

- identify single items in an unstructured pile of similar items, and choose the most favourable item for picking that can be removed and is not obstructed by others;
- take out the single item;
- move the item to a different location; and
- put the item next to or on top of other, non-identical, items.

All this has to be done at a high rate, without damaging (or dropping) the item whether when picking, moving, or placing it in the final location.

The items that have to be picked depend largely on the typical supply chain of which the warehouse is part (e.g. retail, food/non food, pharmaceutical). In many cases, a large variety in size, shape, consistency, and appearance has to be accommodated. Examples of items handled in a retail warehouse in the United Kingdom are shown in Fig. 8.3.

The challenges faced when picking this large variety automatically, fall into two categories, i.e. challenges related to sensing and those related to actuating. For humans, this mainly coincides with vision and grasping. The main challenges are:

Sensing (vision)

- is it possible to identify and locate a candidate item inside the product container?
- is the candidate item free of obstruction by other items in the container?
- where exactly on the item can it be grasped?
- where in the receiving container to place the item?

Fig. 8.3 Samples of the large variety of items to be picked

Actuating (grasping)

- can the item be approached by the gripper?
- how to grasp the item without damaging it?
- has gripping the item been successful?
- is it safe to take the item out of the container towards the receiving container (e.g. without losing it)?
- how can the item be put in the receiving container?

Finding a "one solution suits all" will be very difficult. Such an approach could actually very well end up at the original starting point: a human item picker.

In this part of the book, the approach will be twofold. First, in this chapter, it will be explored what can be achieved by applying simple, "off-the-shelf" technology, for gripping, vision, as well as robotics. Insight is obtained by investigating what such a solution will look like and seeing what functional performance level can be achieved. This is considered as a kind of contemporary *benchmark* in the automation of item picking. Second, in the remaining chapters of this part of the book, alternative solutions to increase the performance of this benchmark will be discussed. A concise preview of these solutions is given at the end of this chapter.

8.3 The Benchmark Solution

Using commercially available, off-the-shelf solutions it will be studied to what extent automation of the item-picking function is feasible. The constituent elements of the overall solution are human-inspired. The elements that will be discussed are on vision, gripping, the arm that carries the gripper, the composition of these into an industrial workstation, and the integrating process control to make the workstation operational.

8.3.1 Vision

There are multiple ways of finding items in an unstructured, random pile of objects in a tote. In industry this class of problems is called *random bin-picking*, solutions of which are manifold as a simple Google query confirms. As the products handled within a distribution centre are known upfront, a visual feature-based approach is possible. A feature is a piece of information that is relevant to fulfil a computational task. Features can be learnt *off-line*, i.e. independently from the task at hand. A clear advantage of this approach is that every item is positively recognised and there is a confirmation of the grab location, the so-called *landing zone* of the gripper.

The feature-based approach is robust against:

- noise (foreign objects, debris, wrong item);
- offset of centre of gravity with respect to geometric centre;
- surface texture suitable for gripping (e.g. porous parts in an item);
- surface shape suitable for gripping (e.g. non-flat surfaces like in a blister);
- vulnerable parts (e.g. flaps that can open when lifting).

The selected open-source software makes use of SIFT keypoints [2] as features. These keypoints define scale-invariant properties of items. An item is detected by finding keypoints in the image that match distinctive keypoints of an item stored in a database. Once the item is identified, all other properties of that item can be related. For instance, the landing zone as defined in the item model (learnt upfront) can be determined in the situation at hand (Fig. 8.4).

8.3.2 Gripping

A commercially available and well-developed area of grippers is using vacuum technology, i.e. suction cups [1]. The use of suction cups does, however, restricts the items to be grasped to items having a relatively rigid, non-porous surface which is flat in the area where one would like to create a vacuum with the suction cup.

Fig. 8.4 Grabbing a product: Landing zones for products

Specific to item picking from totes is that the orientation of the items can vary. In addition, a manipulator and end effector are required to adapt to these orientations. Both issues were addressed by designing a mechanically tolerant gripping head [3], which is shown in Fig. 8.5.

After grasping the item, the item is effectively fixed to the gripper in order to allow the gripper plus item to move to a new position (e.g. the order tote) in a controlled, but high-speed, manner. The deposition of the item in the order tote is controlled in terms of position, but the gripper head as designed does not allow changes of orientation to align with respect to other items in the order tote. This constraint is perceived as a minor disadvantage only.

8.3.3 Arm

In order to move the gripper from the position where the item is picked towards the order tote, some sort of robotic arm is needed. The most commonly known sorts of arms to fulfil this functionality are the articulated robots. These are, however, relatively expensive and the high number of degrees of freedom are not needed in

Fig. 8.5 Flexible suction gripper

pick-and-place movement. The most economical solution is a conventional x, y-gantry robot supporting an end effector able to move in the z-direction and θ-angle.

The dimensions of the arm are chosen such that multiple product totes and order totes can be accessed by the gripper. This allows conventional conveyor belts to supply and take away the totes as well as the supplying system to have ample lead time and non-strict sequence requirements.

8.3.4 Workstation

When composing a workstation from its constituent components, a balance has to be found between the number of items the targeted picker can pick per unit of time, the number of order lines per order (i.e. the number of different items making up a single order) and the average number of items per order line. This leads to different designs depending on the actual application (see also Sect. 8.1).

For the automatic item-picking workstation solution that is composed of elements that were detailed in the previous sections, a picture of part of the workstation is shown in Fig. 8.6. Key characteristics of this workstation are:

- several product totes and order totes are processed simultaneously;
- product and order totes are transported on conveyors;
- items are taken from the product totes and inserted in the order totes while they are on the conveyor.

A movie of the workstation in operation is shown in [4].

Fig. 8.6 Automated item-picking workstation

8.3.5 Process Control

The operational integration of multiple interdependent functions in practice is a challenge. Add to that a process of internal and external development and learning, and things can get complex and entangled.

For the successful integration of functions, a transparent architecture has been chosen (see Fig. 8.7). The baseline is formed by functions having stand-alone functionality with real-time control embedded. The building blocks are for example *find an item in a scene* and *move the gripper from A to B*. These stand-alone functions have no knowledge of their surroundings and can be addressed as services.

The control of the item-picking process is done by a conductor-like function called *process control*. This function has knowledge of the configuration of the workstation and the required steps to accomplish the item-picking task at hand. This conductor addresses the services that are available by the underlying functions. When to use which service (scheduling) is a key functionality of the process controller (see Fig. 8.8).

Fig. 8.7 Modular approach of system architecture

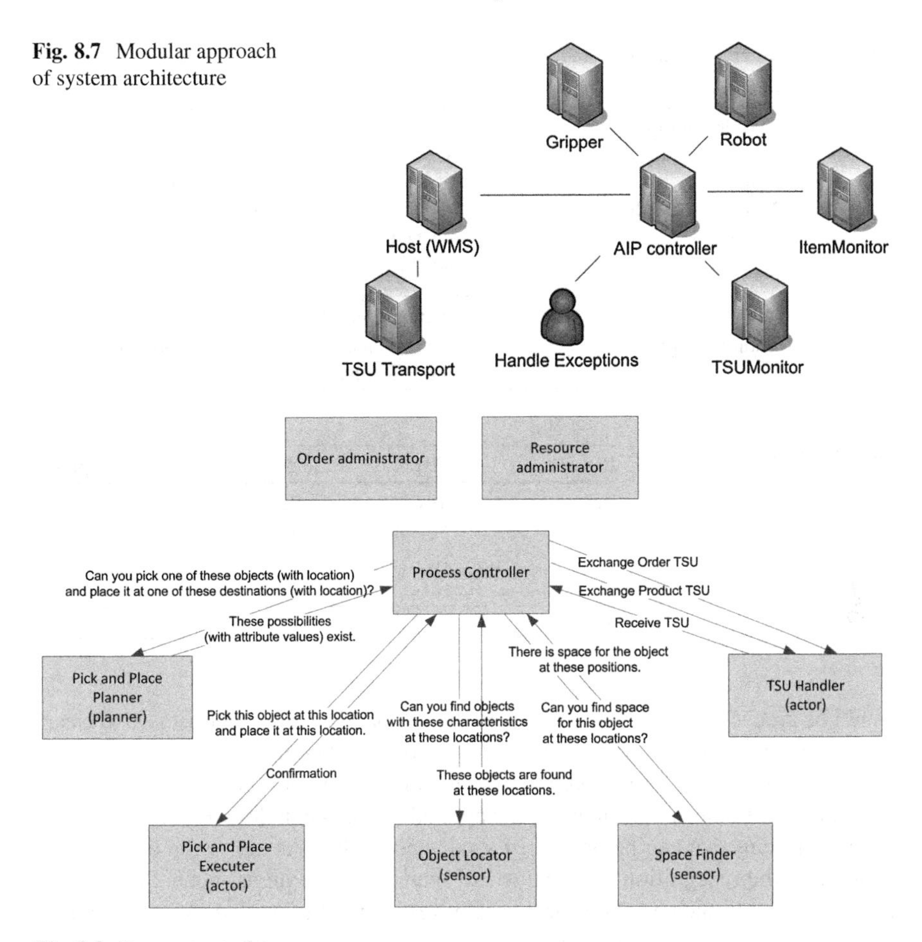

Fig. 8.8 Process control

8.4 The Working Range of the Benchmark Solution

The choice for relatively standard components instigates obvious limitations as to what the designed workstation can achieve functionally. These limitations are listed in Table 8.1.

The relevance of impact from these restrictions can be assessed from the data obtained from a retail, non-food distribution centre logging the items picked in one week. In total more than 600,000 picks were analysed. All items were assessed on among others surface rigidity and the availability of a flat area for suction (see Table 8.2).

As can be seen from Table 8.2, 72% of the items, or 64% of the number of picks, are available to be picked by the automatic item-picking workstation. This leaves approximately 28% of the items, accounting for 36% of the number of picks, that

Table 8.1 Limitations of the workstation

	Solution	Restrictions	Will not work
Gripper	Flexible, multiple vacuum cup suction head	Rigid, flat, non-permeable surfaces	Soft, permeable, uneven or very small surfaces
Arm	Gantry robot with z-axis applicator	Items approached/retrieved upwards; Items placed in same orientation as picked	Items stored in unsuitable directions
Vision	Feature recognition	Item needs sufficient and stable features	Featureless or shiny (e.g. plastic) surfaces

Table 8.2 Item rigidity and flatness characteristics

Rigid	Flat	% items	% picks
Yes	Yes	72	64
No	Yes	4	4
Yes	No	20	24
No	No	4	8
Total		100	100

cannot be processed in the workstation by the gripper. From the table, it is also clear that the major part of the items that cannot be processed do have a rigid surface, but do not have a flat area for the suction cups to contact the item properly.

In addition to the above analysis, the applied vision technology restricts the percentage of items that can be processed somewhat further. As the vision technology is based on the recognition and analysis of features of the items, items that do not have any features such as plain cardboard boxes cannot be handled. Also, when the surface material of the item is very shiny, reflections of the illumination will make the features unrecognisable for the vision system.

Figure 8.9 shows some examples of items that cannot be handled by the automatic workstation described in this chapter.

8.5 Conclusion and Outlook

An industrial item-picking workstation can handle many of the items that are typical to a warehouse environment. However, it cannot handle all items. These are items that either do not have a rigid surface or flat areas for suction to work, the latter being a major group of items typical for a retail, non-food warehouse. A gripper that would be able to grasp all of the remaining items would be ideal. A fully controlled and actuated robotic hand would be an option, however, price and robustness are not yet at a level that appears to be practical for industrial application in a warehouse environment. An interesting solution is the use of so-called underactuated grippers. This will be discussed in Chap. 9.

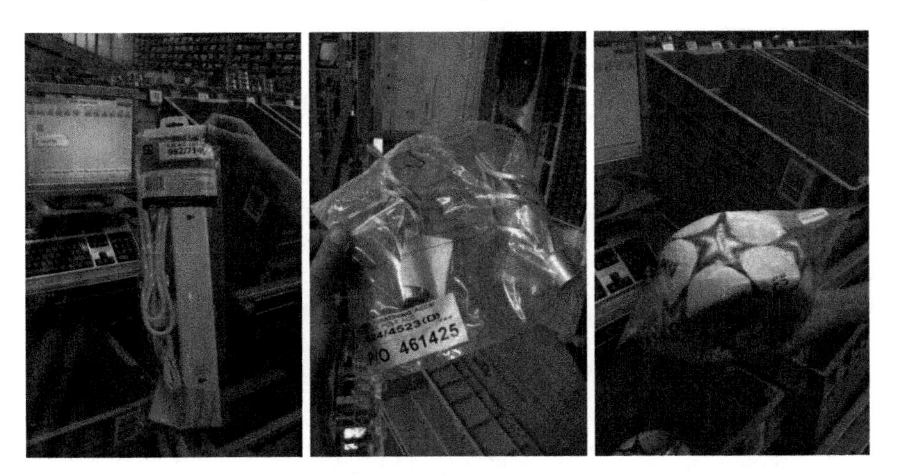

Fig. 8.9 "Problematic items": soft, not flat, shiny, no features, changing appearance

From a vision point of view, in order to further increase the percentage of items that can be handled by the vision system, especially those items that do not have any features need to be characterised differently. Improved item characterisation is discussed in Chaps. 10 and 11, where separate attention is devoted to learning the representation of items in a database and the recognition task itself.

Lastly, the integration of alternative approaches to gripping and vision is discussed in Chap. 12. It is shown that significantly better solutions are indeed feasible at competitive costs. Nevertheless, complete coverage of all pickable items appears difficult to achieve in a robust way. A solution to deal with this imperfection of automated item picking has to be constructed at the system level by introducing hybrid solutions.

References

1. Festo (2011) Vacuum suction grippers and suction cups. http://www.festo.com, Viewed April 2011
2. Lowe DG (2004) Distinctive image features from scale-invariant keypoints. Int J Comp Vis 60:91–110
3. Vanderlande Industries (2009) Inrichting voor het grijpen van objecten. Patent application NL1037512
4. Vanderlande Industries (2010) Automated item picking. http://www.youtube.com/watch?v=4hjmLYvy5DI, Viewed April 2011

Chapter 9
Underactuated Robotic Hands for Grasping in Warehouses

Gert Kragten, Frans van der Helm, and Just Herder

Abstract The automation of order picking in warehouses requires adaptive robotic hands that are cheap, robust, easy to control, and capable to reliably grasp a large range of products. The state-of-the-art graspers or robotic hands cannot fulfil this need. The goal of this research is to design and evaluate an adaptive and simple hand by applying the concept of *underactuation* (i.e. having fewer actuators than independently moving fingers). The innovation is that this hand mechanically decides to firmly envelope large objects, and to grasp small objects between the tips of the fingers. Experiments with the new hand show that the full range of object sizes that appear in the workstation of a reference warehouse can be grasped with a minimum control effort and electronic parts (i.e. one motor and no sensors). It is concluded that this new, underactuated hand is applicable for automated order picking.

9.1 Introduction

Order picking by human operators is a limiting factor to increase the profit and the throughput of warehouses in the future (see Chap. 1). Automation of this process is needed, which requires grippers that are cheap, robust, easy to control, and capable of reliably grasping all common products in the warehouse. However, current industrial solutions cannot fulfil this need (see Chap. 8). State-of-the-art robotic hands are also

G. Kragten (✉) · F. van der Helm · J. Herder
Faculty of Mechanical, Maritime and Materials Engineering,
Delft University of Technology, Mekelweg 2, 2628 CD Delft, The Netherlands
e-mail: g.a.kragten@gmail.com

F. van der Helm
e-mail: f.c.t.vanderhelm@tudelft.nl

J. Herder
e-mail: j.l.herder@tudelft.nl

R. Hamberg and J. Verriet (eds.), *Automation in Warehouse Development*,
DOI: 10.1007/978-0-85729-968-0_9, © Springer-Verlag London Limited 2012

not suited. These hands typically consist of three or four fingers with two or three phalanges per finger. As every phalanx is independently actuated and controlled, the number of expensive and vulnerable electronic components is too large, and the control architecture is too complex to be able to quickly and reliably grasp thousands of different products per day. Thus, a totally different design approach is needed to obtain dexterous hands that satisfy the industrial needs.

Underactuation is a promising concept to yield cheap, robust and reliably grasping hands. Underactuated hands have fewer actuators than independently moving fingers and phalanges. A reduction of actuators means less weight, less costs, and a more robust hand. Furthermore, this reduction in actuators provides the fingers the capability to adapt themselves to the objects such that easy and safe grasping of objects of unknown and irregular shapes is possible. It is assumed that underactuated hands can satisfy the need for automated order picking.

A number of prototypes of underactuated hands have been presented in the literature (e.g. Birglen et al. [1]). These prototypes easily grasp objects of various size, shape, and weight in a so-called *power grasp* where the fingers envelope the objects. However, the performance of these hands has never been unambiguously defined or quantified. The effect of mechanical design choices on the performance is unknown. Furthermore, grasping small or slender objects in a so-called *precision grasp*—where the objects are grasped by only the distal phalanges—is unreliable because of the instability of the fingers and object in this configuration. Current solutions to grasp small objects by underactuated fingers are complex and expensive. Quantification and improvement of the performance is needed to let underactuated hands successfully pick orders in warehouses.

The objectives of this research are as follows:

1. Define functional performance metrics for underactuated hands and quantified design requirements for application in warehouses.
2. Design and evaluate a new hand that can grasp all relevant object sizes and that suits to the requirements of the warehouse.

9.2 Performance Metrics and Design Requirements

Operators at an order-picking workstation use their hands mainly for the following three functions:

1. Grasp one or a few objects lying in a product tote.
2. Hold the objects while they move to the order tote.
3. Release the objects in the order tote.

There is a large variation in size, shape, and weight of the objects. Data analysis of the entire product set in a reference warehouse yielded the following characteristics to the size and weight. 95% of the products has a minimum cross section that is smaller than 118 mm and a mass smaller than 1.1 kg. Approximately 75% of the products has a box shape, while 25% has an irregular shape. Assuming that rectangular objects are going to be grasped by a suction device (see Chap. 8), a robotic hand must be able to grasp, hold, and release these irregular products in order to automate the workstation.

Grasping objects means achieving force and moment equilibrium of the hand with the object. *Holding* objects is maintaining the grasp equilibrium while external forces apply that are caused by gravity and accelerations. *Releasing* objects involves loosing contact between the hand and the object, while the object is in the desired orientation and position. It is assumed that orientating and positioning of the objects is achieved by a robot arm. The performance of underactuated hands to grasp and hold objects is unknown.

We propose two performance metrics to address the *ability to grasp* different sizes of objects and the *ability to hold* the objects when force disturbances apply. The first metric determines the smallest and largest diameter of cylindrical objects that can be successfully grasped such that fingers and object are in equilibrium. The second metric quantifies the maximum allowable force which can be applied to a grasped object before it escapes from the grasp. The definition of the performance metrics with cylindrical objects allows calculation by a simple mathematical model for analysis and design optimisation. It also allows evaluation by reproducible experiments. Kragten et al. [3, 4, 6] provide a detailed elaboration of the mathematical model and the experimental method and materials.

Design requirements are defined based on the size and weight of the objects in the data set of the reference warehouse and in accordance with the defined performance metrics. The following performance and properties are required:

- Achieve grasp equilibrium for objects with a diameter up to 120 mm. These objects initially lie on a flat surface and are approached from above by the hand.
- Mechanically decide to grasp small objects in a precision grasp and large objects in a power grasp. There is no additional mechanism, motor, or feedback system to achieve precision grasps.
- Hold objects in a power grasp at force disturbances up to 20 N, and hold objects in a precision grasp at force disturbances up to 10 N.
- The time to fully close or open the hand is less than 0.5 s.
- The hand has a minimum number of electronic components. The torque of the motor is distributed by conventional gears and linkages to achieve a cost-effective and robust hand.

9.3 Design and Evaluation of the Delft Hand 3

A new robotic hand was designed for order picking of irregularly shaped products in a fully automated warehouse. A picture of a prototype of this new hand is shown in Fig. 9.1. It consists of three identical fingers of two phalanges each. All six phalanges can independently rotate, while they are driven by a single DC-motor. The main innovation is that the decision to grasp large objects in a power grasp (i.e. the fingers envelope the object) and small objects in a precision grasp (i.e. the object is

Fig. 9.1 Photo of the Delft Hand 3

grasped by the tips of the fingers) is fully incorporated in the mechanical hardware. Active switching between the two grasp types by a control system is not needed. The following sections address the conceptual design, dimensional design approach, and performance evaluation of this hand. More details are provided by Kragten et al. [8].

9.3.1 Conceptual Design

The concept to drive the fingers of the Delft Hand 3 is based on a previous prototype [9]. The torque at the shaft of the motor is equally divided by a differential mechanism and drives the two fingers at one side of the palm. The reaction torque at the housing of the motor drives the single finger at the opposite side of the palm. The torque applied to the fingers is distributed to the phalanges through a four-bar linkage mechanism (see the schematic side view of the hand in Fig. 9.2). This mechanism lets a finger rotate as a single body about the proximal joint J_1 when there is no contact with the object. The distal joint J_2 remains against an extension stop due to the spring between the proximal phalanx and link B. When the proximal phalanx gets in contact with an object, the mechanism lets the distal phalanx continue to rotate about joint J_2 until contact between the object and this phalanx is established. From a force perspective, the mechanism applies torques T_1 about joint J_1 and T_2 about joint J_2 (see the static equivalence diagram in Fig. 9.3). T_1 is equivalent to the torque applied to link A by the motor. T_2 is equivalent to $R(\theta_2)T_1$, where $R(\theta_2)$ is the *transmission ratio* between input link A and output link C of the four-bar mechanism. This ratio is a function of the rotation θ_2 of the distal phalanx, the dimensions of link B, C, L_1 relative to link A, and the properties of the return spring.

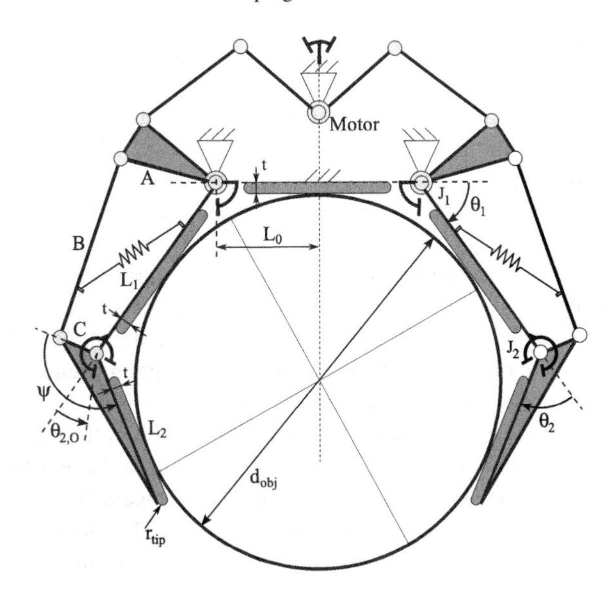

Fig. 9.2 Schematic side view of the Delft Hand 3 with the fingers facing down. The *right side* shows the proximal and distal joint J_1 and J_2, respectively, and their rotation angle θ_1 and θ_2. The *left side* shows the symbols to denote the width of the palm L_0, the length of the proximal phalanx L_1 and the distal phalanx L_2. t is the thickness of the fingers and palm. $\theta_{2,O}$ is the minimum rotation angle of the distal phalanx, limited by an extension stop. r_{tip} is the radius at the curved top of the fingers. The object with a diameter d_{obj} is grasped in a power grasp configuration with contact at each phalanx and the palm

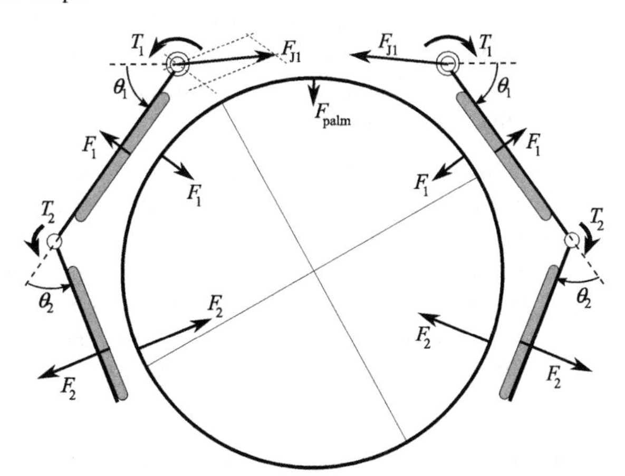

Fig. 9.3 Static equivalence diagram of the fingers and object in a power grasp configuration. T_1 and T_2 are the equivalent actuation torques applied by the actuation mechanism to the proximal and distal phalanges, respectively. F_1 and F_2 are the contact forces between the object and proximal and distal phalanges, respectively. F_{palm} is the contact force between the object and palm. F_{J1} is the radial force in the proximal joints

The concept to let the fingers mechanically decide between a power grasp and precision grasp is as follows. When the initial contact between the fingers and the object is at the curved tip, the distal phalanx remains against its extension stop and the finger acts as a rigid body that rotates about joint J_1. The distal phalanx does not rotate, because the four-bar linkage is designed such that the actuation torque T_2 about J_2 is smaller than the counteracting moment about J_2 caused by a contact force at the tip. Note that the magnitude of this force is equal to T_1 times the moment arm between J_1 and the contact point at the tip. Hence, a precision grasp between the tips of the fingers is established. However, when the initial contact is not at the tip, the distal phalanx will flex because the counteracting moment about J_2 is relatively smaller than T_2. The object is then pushed towards the palm. Hence, a power grasp where the object is enveloped by the palm and the fingers is established.

The dimensioning of all links, phalanges and the width of the palm is a trade-off where on one side the fingers can envelope objects with a diameter up to 120 mm and T_2 is large enough to establish equilibrium of the fingers and object in a power grasp. On the other side, the fingers also have to pick small objects from the bottom of a tote where T_2 is small enough to establish equilibrium of the fingers and object in a precision grasp.

9.3.2 Dimensional Design Approach

The dimensions of the phalanges of the fingers (L_1, L_2), the width of the palm (L_0), the linkages of the actuation mechanism (B, C, and angle ψ) and the rotation limits of the phalanges are determined by an innovative, model-based design approach. The symbols are visualised in Figs. 9.2 and 9.3 and explained in Table 9.1. This approach passes through three subsequent design steps:

1. Step 1 determines constraints on L_0, L_1, L_2, and the transmission ratio $R(\theta_2)$, such that objects with a diameter between $2L_0$ and 120 mm can be grasped and held in a power grasp configuration.
2. Step 2 determines L_0, L_1, L_2, $\theta_{2,O}$, and $R(\theta_{2,O})$, such that objects with a diameter smaller than $2L_0$ can be grasped in a precision grasp from the bottom of a tote, while conflicts with the constraints of step 1 are prevented.
3. Step 3 determines B, C, and angle ψ, such that the constraints on $R(\theta_2)$ are satisfied.

Note that several dimensions of the hand (e.g. the total length L of the fingers) are chosen beforehand. These dimensions are separately shown in Table 9.1.

9.3.2.1 Design step 1 : Grasping large objects

The maximum diameter $d_{\mathrm{obj,\,max}}$ of a cylindrical object that can be enveloped is determined by geometric equations. Enveloping means here that $\theta_1 + \theta_2 > \pi/2$, and

Table 9.1 Symbols and explanation of the main design variables and chosen dimensions

Symbol	Dimension	Explanation
Main predefined dimensions		
L	100 (mm)	Total length of the finger
A	20.0 (mm)	Length of link A of the actuation mechanism
T_1	1.0 (Nm)	Motor torque and torque applied to link A
Main design variables		
L_0	28.0 (mm)	Palm width
L_1	56.0 (mm)	Length of the proximal phalanges
L_2	38.8 (mm)	Length of the distal phalanges (excluding the curved tip)
$\theta_{2,O}$	$\pi/6$ (rad)	Minimal rotation angle of the distal phalanges
B	50.5 (mm)	Length of link B of the actuation mechanism
C	9.40 (mm)	Length of link C of the actuation mechanism
ψ	2.30 (rad)	Angle between link C and the distal phalanx

Most of the symbols are visualised in Fig 9.2. The dimensions of the main variables are determined by three subsequent design steps

that the palm and each phalanx is in contact with the object. It is assumed that the object is on the symmetry line as drawn in Fig. 9.2. The geometric equations to determine $d_{\text{obj, max}}$ are further elaborated by Kragten and Herder [5]. The result is summarised by the following equation:

$$d_{\text{obj, max}} = L_1 + \sqrt{L_1^2 - 4L_0^2 + 4L_0L_1} - 2t. \tag{9.1}$$

The maximum diameter $d_{\text{obj, max}}$ is visualised by contour lines in Fig. 9.4 as a function of L_0 and L_1, where it is assumed that $L = 100$ mm, $t = 3$ mm, and $L_2 = L - L_1 - 6$ mm. The contour line at $d_{\text{obj, max}} = 120$ mm, is emphasised, as objects up to this size have to be grasped according to the design requirements. This means that only dimensions of L_0 and L_1 above this contour line and between the dashed lines can be chosen in order to satisfy this design requirement. It is decided that $L_0 = 0.5L_1$, because this results in the largest $d_{\text{obj, max}}$ at a particular choice of L_1 (as long as L_1 is smaller than 62.7 mm). Substitution of $L_0 = 0.5L_1, t = 3$, and $d_{\text{obj, max}} = 120$ into Eq. 9.1 yields that L_1 has to be larger than 52.2 mm.

The fact that the fingers can envelope objects with a diameter up to 120 mm does not directly imply that equilibrium of the fingers and object exists at this configuration. The fingers are in equilibrium in the power grasp configuration if the contact forces F_1 and F_2 are positive (i.e. in the direction as drawn in Fig. 9.3, because only compressive contact forces are feasible). The object is in equilibrium if the resultant of the contact forces F_1, F_2, and F_{palm} is zero, and all forces are positive. The magnitude of F_1, F_2, and F_{palm} is given by the following equations:

$$F_1 = \frac{T_1 - RT_1 - F_2L_1\cos\theta_2}{p_1}, \tag{9.2}$$

$$F_2 = \frac{RT_1}{p_2}, \tag{9.3}$$

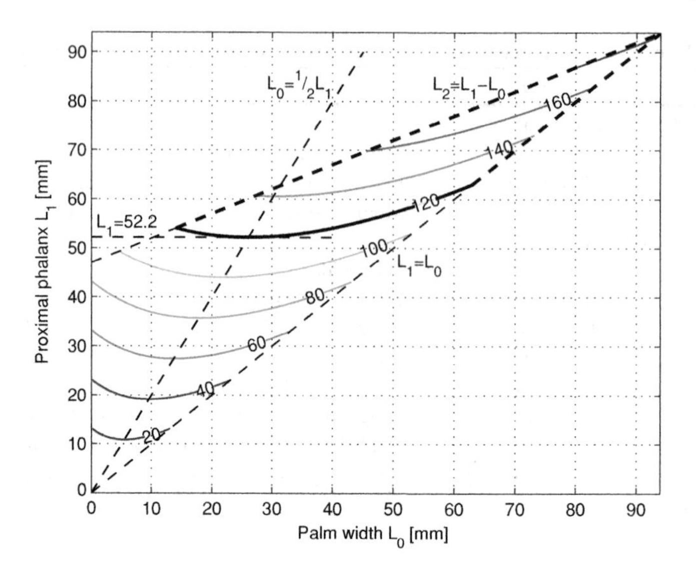

Fig. 9.4 *Contour lines* showing the maximum object diameter that can be enveloped by the fingers as a function of the palm width and the proximal phalanx length. It is assumed that the distal phalanx length is equal to $L_2 = L - L_1 - 6$, where L is the total finger length and 6 is subtracted to account for the curved tip. To grasp objects larger than the required $d_{\text{obj}} = 120$ in a power grasp configuration, the dimensions of L_0 and L_1 have to satisfy the design space above the 120 *contour line* and between the *dashed lines*

$$F_{\text{palm}} = 2\left(-F_1 \cos\theta_1 - F_2 \cos(\theta_1 + \theta_2)\right), \tag{9.4}$$

where p_1 and p_2 are the contact point positions on the proximal and distal phalanx, respectively. The rotation angles θ_1 and θ_2 depend on the object size. It was elaborated by Kragten and Herder [5] that the constraints $F_1 > 0$ and $F_{\text{palm}} > 0$ yield two constraints on the transmission ratio $R(\theta_2)$. The result is summarised by the following equation:

$$\frac{2\left(M - L_0^2\right)\left(M + (L_0 - L_1)^2\right)}{4M\left(L_0^2 + M\right)} \leq R \leq \frac{(L_0 - L_1)\left(M + (L_0 - L_1)^2\right)}{M\,(L_0 - 2L_1) + L_0\,(L_0 - L_1)^2}, \tag{9.5}$$

where $M = \frac{1}{4}(d_{\text{obj}} - 2t)^2$. Note that there is a non-linear relation between d_{obj} and the rotation angle θ_2 when the distal phalanx is in contact with the object in the power grasp configuration. The constraints on R as a function of d_{obj} and θ_2 are visualised in Fig. 9.5, where $L_0 = 28$ and $L_1 = 56$ mm are substituted into Eq. 9.5. It visualises, for example, that $0.40 < R(\theta_2 = 0.84) < 0.43$ to achieve equilibrium of the fingers and an object of $d_{\text{obj}} = 120$ mm. Indeed, these constraints on R depend on the dimensions of L_0 and L_1.

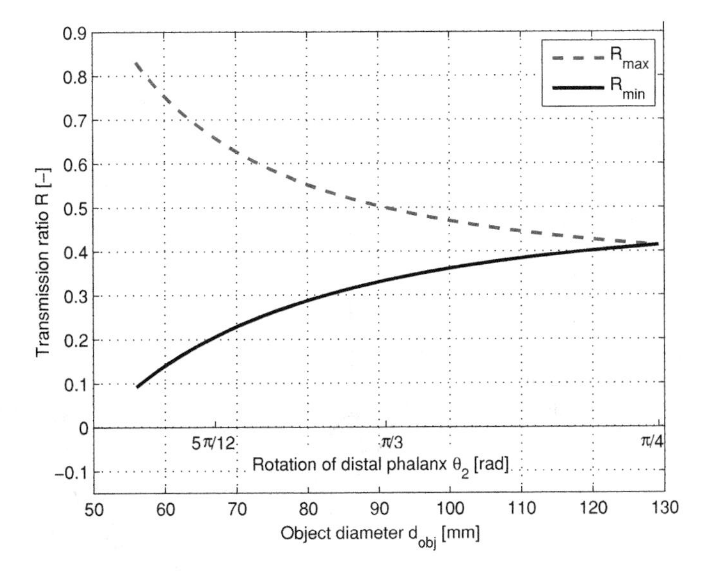

Fig. 9.5 Minimum and maximum transmission ratio R as a function of the diameter of the object in order to achieve equilibrium of the fingers and object in a power grasp configuration ($L_0 = 28$, $L_1 = 56$ mm). A horizontal axis with the rotation angle of the distal phalanx is superposed. R is the ratio of the actuation torque applied to the distal phalanx relative to the proximal phalanx, see Fig. 9.3. This ratio is critical to grasp large objects. At $d_{obj} = 129$ mm ($\theta_2 = \pi/4$), the lines intersect at $R = 0.414$. Larger objects cannot be grasped at the selected size of L_0 and L_1

In summary, design step 1 led to the following design decisions and constraints: $L_0 = 0.5L_1$, $L_1 > 52.2$ mm, $L_2 > 94 - L_1$ mm, and $R(\theta_2)$ within the bounds determined by Eq. 9.5. Based on experience with a previous prototype, it is assumed that an actuation torque of $T_1 = 1$ Nm is sufficient to maintain the power grasp while external forces of 20 N or more apply. More details to calculate the force disturbance rejection are provided by Kragten and Herder [4].

9.3.2.2 Design step 2: Grasping small objects

The extension stop of the distal phalanx with respect to the proximal phalanx is an effective design parameter to easily achieve equilibrium of underactuated fingers in precision grasp configurations [2]. It was explained in Sect. 9.1 that the fingers are in equilibrium in a precision grasp configuration if the actuation torque T_2 about the distal joint J_2 is smaller than the counteracting moment about J_2 caused by the contact force at the tip. This leads to a constraint on the transmission ratio R when $\theta_2 = \theta_{2,O}$. Before this constraint is determined, first geometric constraints have to be determined such that sufficiently small objects can be touched by the tips of the fingers when they lie at the bottom of a tote. Furthermore, the friction coefficient

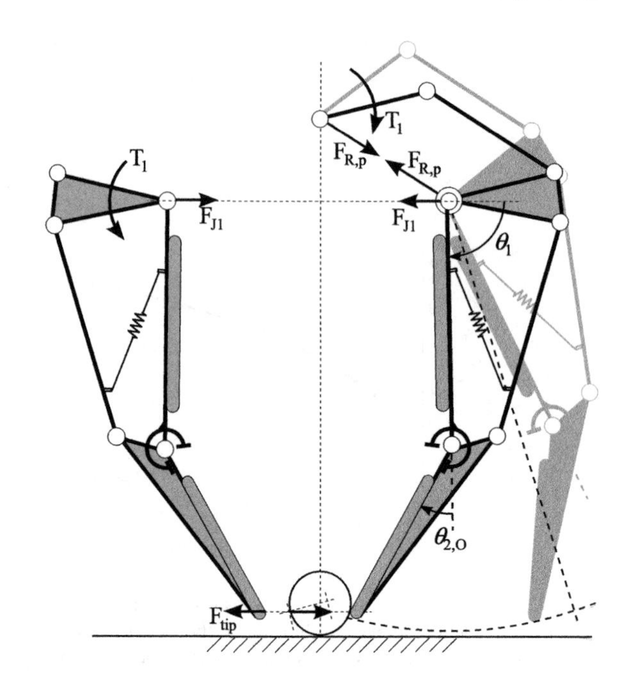

Fig. 9.6 Schematic drawing of a precision grasp. The *left side* shows the contact force F_{tip} at the contact point, whose magnitude is determined by the actuation torque T_1 and the moment arm between the contact point and the proximal joint. The *right side* shows the rotation of the finger until it touches the object. *Construction lines* are drawn to show that collision with the ground has to be prevented, and that the minimum object size is limited by either the geometry of the hand or the maximum tangential tip force

between the objects and finger tips has to be sufficient to achieve equilibrium of the object.

The geometric equations and the force equilibrium equations are derived by Fig. 9.6, and elaborated in more detail by Kragten et al. [8]. It is assumed that the radius of the tip $r_{\text{tip}} = 6$ mm and that a friction coefficient $\mu = 0.5$ can be achieved between the object and the finger tips. The size of the smallest object that can be grasped is visualised in Fig. 9.7 as a function of L_1 and $\theta_{2,O}$ (and $L_0 = 0.5L_1$). The constraint on the transmission ratio $R(\theta_2 = \theta_{2,O})$ is visualised in Fig. 9.7, where R has to be smaller than the values shown by the contour lines to achieve equilibrium of the fingers.

Based on Fig. 9.7 and the constraints from the previous design step, it is decided that $L_1 = 56$ mm and $\theta_{2,O} = \pi/6$ rad. At this point, the size of objects that can be grasped in a precision grasp is near the minimum (i.e. 7.5 mm and larger). It is supposed that the constraint on the transmission ratio, $R(\theta_2 = \pi/6) < 0.38$, does not cause conflicts with the constraints from the previous design step. Note that the weight of the object is not taken into account, meaning that the actual minimal object size will be larger.

According to the design requirements, the precision grasp has to be maintained while external forces of 10 N apply to the object. It is assumed that this is satisfied when the fingers apply a normal contact force of 10 N while the friction coefficient $\mu = 0.5$. The moment arm between the proximal joint J_1 and the finger tip is approximately 100 mm. Hence, an actuation torque of $T_1 = 1$ Nm satisfies.

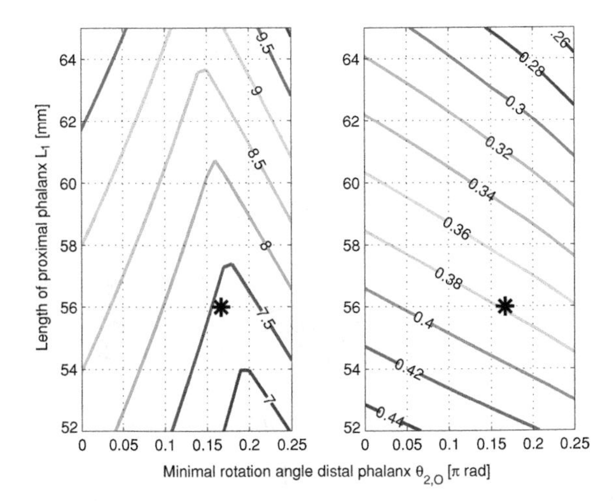

Fig. 9.7 *Contour lines* in the *left* figure show the minimal object size that can be grasped from a flat surface into a precision grasp as a function of the minimal rotation angle of the distal phalanx $\theta_{2,O}$ and the length of the proximal phalanx L_1 ($L_0 = 0.5L_1$). The asterisk shows the chosen dimensions. The *contour lines* in the *right* figure show the maximal transmission ratio R as a function of $\theta_{2,O}$ and L_1, such that the distal phalanx remains against the extension stop while objects are grasped in a precision grasp configuration

9.3.2.3 Design step 3 : Dimensioning of the actuation mechanism

The previous design steps defined constraints on the transmission ratio R of the actuation mechanism at various rotation angles of the distal phalanx. Various implementations of underactuated mechanisms to drive the phalanges exist, for example with cables and pulleys, cams, or with multiple linkages. Each type of mechanism offers a number of design variables and limitations in order to satisfy the constraints. A four-bar linkage was chosen for the Delft Hand 3, because it consists of a relatively small number of links, and it can be constructed by robust elements. In addition, it has a non-linear transmission ratio which is needed to satisfy the design constraints determined by the previous steps.

The length of link B, C, angle ψ, and the attachment points of the return spring are design variables of the actuation mechanism of the Delft Hand 3. The length of link A and the properties of the spring were chosen beforehand. These dimensions were determined by a non-linear optimisation algorithm within the following constraints: $R(\pi/6) = 0.38$, $R(\pi/4) = 0.41$. The objective function was to maximise $R(\pi/2)$.

9.3.3 Performance Evaluation

The performance to grasp and hold objects by the Delft Hand 3 was experimentally evaluated. Benchmark tests, as proposed by Kragten et al. [6], were performed to

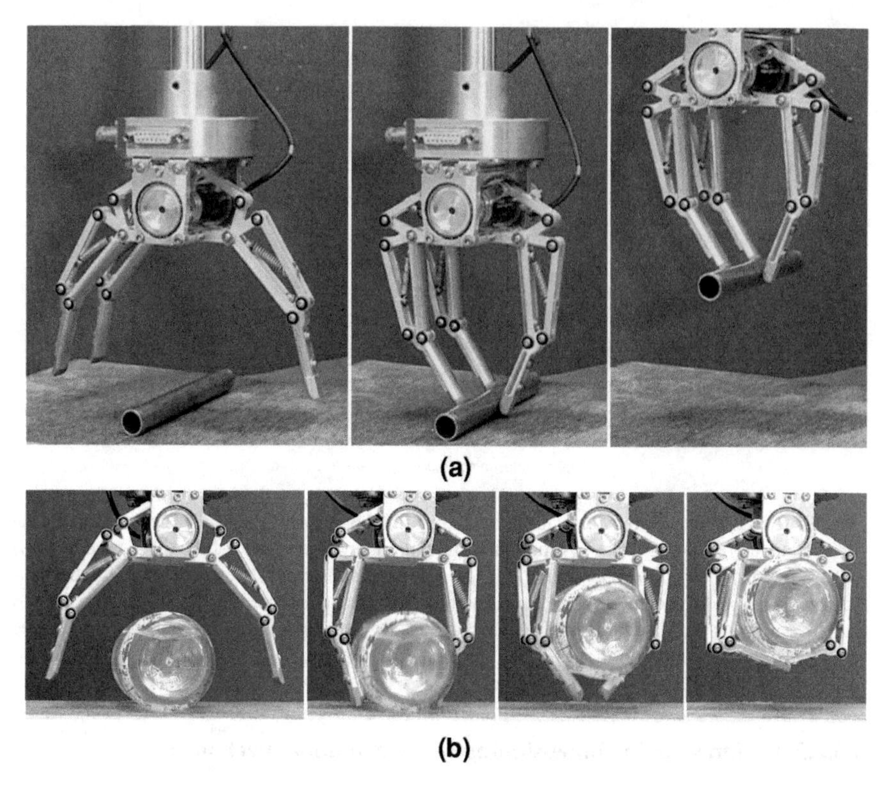

Fig. 9.8 Snapshots of grasping an object from a flat surface. **a** shows a precision grasp of an object with a diameter of 14 mm. **b** shows a reconfiguration to a power grasp of an object with a diameter of 60 mm

quantify the size of objects that the hand can grasp (i.e. ability to grasp) and the maximum disturbance force that can be applied to the grasped objects (i.e. ability to hold). The ability to grasp objects was tested with the hand mounted on a robot arm with the fingers facing down. Cylindrical objects with diameters between 10 and 120 mm were lying on a flat surface. It was observed whether grasp equilibrium could be achieved after the hand was closed and lifted straight upward, see Fig. 9.8. Successful grasping of objects with a diameter of 14 mm and larger was observed in this way. For large objects, it was necessary that the palm came close to the objects before closing the fingers.

The ability to hold objects was tested on a tensile test bench, see Fig. 9.9. Objects with a diameter of 10, 30, 55, 76, 90, 105, and 120 mm were pulled all the way out of the hand in the direction perpendicular to the palm until the fingers lost contact. The force was measured to determine the maximum external force that had to be applied to the object to achieve full ejection. The results are shown in Table 9.2. Note that the object with a diameter of 55 mm was disturbed from a power grasp configuration as well as a precision grasp configuration. The maximum external load on objects in

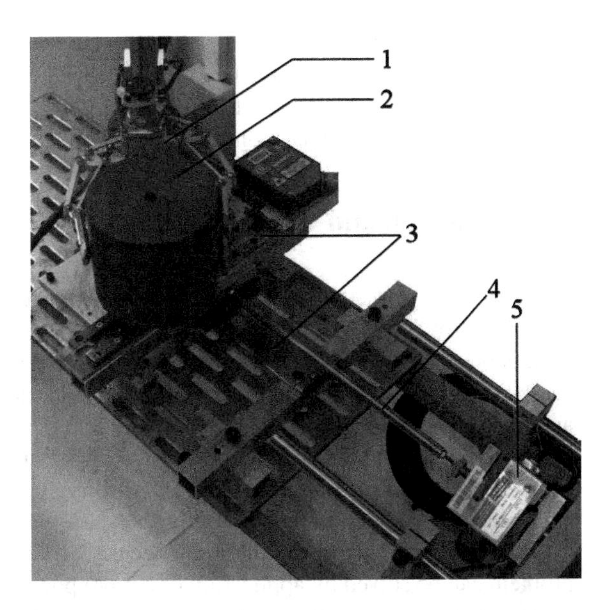

Fig. 9.9 Experimental setup where *1* is the Delft Hand 3; *2* is an object; *3* are linear bearings to let the object move in a plane; *4* is a rod to pull the object; and *5* is the force sensor

Table 9.2 Maximum external force [N] that was applied to objects of a diameter between 10 and 120 mm to pull them all the way out of the hand

Diameter (mm)	Force (N)	
	Precision grasp	Power grasp
10	4.1 (4.0, 5.3)	–
30	4.2 (3.7, 5.3)	–
55	5.3 (4.7, 5.8)	156 (152, 159)
76	–	63.7 (61.6, 66.3)
90	–	40.3 (40.0, 41.2)
105	–	23.3 (22.0, 24.5)
120	–	4.9 (4.7, 9.0)

The median values of five repetitions are shown, while the minimum and maximum measured value is shown between brackets. The middle column belongs to objects grasped in a precision grasp, whereas the right column belongs to the objects grasped in a power grasp configuration

a precision grasp is about 4 N, which is smaller than the required 10 N. Objects in a power grasp configuration with a size up to 105 mm need more than the required 20 N to be ejected. Less force is needed for larger objects. It was observed that the efficiency of the gear of the motor caused a smaller actuation torque than expected. Consequently, the contact forces were smaller and the requirements to hold objects when disturbance forces apply were not fully satisfied.

The Delft Hand 3 was also tested for its capability to grasp a set of products that are representative for a warehouse. It successfully grasped products with a large variety of shapes, sizes, compliance, and weight [7]. The closing and opening time of the hand is less than 0.5 s. The weight of the hand including the motor is 330 gram. The total production costs of the first prototype were less than 2.5 k€. It is concluded that this new underactuated hand satisfies the industrial need of an adaptive, robust, and cost-effective hand for automatic grasping in a warehouse.

9.4 Discussion and Conclusion

The goal of this research was primarily to design and evaluate an adaptive, robust, and simple robotic hand for application in an automated order-picking workstation of a warehouse. Experiments with the new hand showed that a large variety of products can be grasped with a minimum control effort and electronic parts (i.e. one motor and no sensors). Grasp equilibrium exists for the full range of object sizes that appear in the workstation of the reference warehouse. The hand decides itself to grasp small and slender objects in a precision grasp and large objects in a power grasp. The capability to maintain grasp equilibrium when force disturbances apply is weaker than required due to the disappointing efficiency of the gear of the current motor. This problem can be solved by application of a more powerful or more efficient motor.

The Delft Hand 3 allows easy integration in a fully automated workstation, such as the one described in Chap. 12. The control effort of the hand is minimal (i.e. binary). Precise positioning of the hand with respect to the products is not necessary, because the fingers adapt when the object is off-centred. In addition, the fingers are compliant at collision. The fingers and palm are slender which allows easy integration with vision or even a suction gripper. Suction cups can be mounted, for example, at the outside of the distal phalanges. When the fingers are fully flexed, these cups are facing down and flat products can grasped. It is concluded that the Delft Hand 3 satisfies the industrial need.

This research led to results that are generally applicable for the design and evaluation of dexterous or underactuated hands. Two of them are highlighted in this discussion. The first result is the new performance metrics that quantify the ability to grasp and hold objects. The application of cylindrical objects simplifies the calculations needed to substantiate the design choices, and it allows easily reproducible experiments to evaluate the performance of the hand. The assumption of cylindrical objects is reliable, although not formally validated. The Delft Hand 3 (and previous prototypes) is able to grasp objects with all kind of shapes with a cross section similar to the calculated and measured diameter of the cylindrical objects. It is concluded that these performance metrics and related experiments can be applied as benchmark tests to assess and compare the performance of robotic hands in general.

The second generally applicable result is the approach to design underactuated hands. This approach is characterised by steps that first determine the dimensions of the phalanges and palm in order to grasp the required range of object sizes in a power or precision configuration. Figures 9.4, 9.5, and 9.7 can be applied as design charts to select the dimensions of the phalanges and palm and to define constraints on the transmission ratio of the mechanism that distributes the motor torque to the phalanges. These constraints are necessary to obtain stable grasp equilibrium of the fingers and object in the power and precision configuration. This is followed by the selection of a conceptual solution to actuate the phalanges (e.g. with pulleys or linkages) and calculation of its dimensions such that the constraints are satisfied. This approach can be repeated when adaptive underactuated hands are needed for warehouses or applications where the size of the products is different. It is concluded that the stepwise design approach is generally applicable to design underactuated hands with the optimal dimensions to grasp and hold the required range of object sizes in the desired configuration.

References

1. Birglen L, Laliberté T, Gosselin C (2008) Underactuated robotic hands, Springer tracts in advanced robotics vol 40. Springer, Heidelberg
2. Kragten GA, Baril M, Gosselin C, Herder JL (2011) Stable precision grasps by underactuated grippers. IEEE Trans on Rob (in press)
3. Kragten GA, Herder JL (2010) The ability of underactuated hands to grasp and hold objects. Mech Mach Theory 45:408–425
4. Kragten GA, Herder JL (2011) An energy approach to analyze and optimize grasp stability in underactuated hands
5. Kragten GA, Herder JL, van der Helm FCT (2011) A planar geometric design approach for a large grasp range in underactuated hands. Mech Mach Theory 46:1121–1136
6. Kragten GA, Meijneke C, Herder JL (2010) A proposal for benchmark tests for underactuated or compliant hands. Mech Sci 1:13–18
7. Kragten GA, Meijneke C, Herder JL (2011) Delft Hand 3: An adaptive, underactuated hand for power and precision grasps. http://www.youtube.com/watch?v=AHoFSuSEe6k, Viewed April 2011
8. Kragten GA, Meijneke C, Herder JL (2011) Design and evaluation of an underactuated hand that mechanically decides between a power and precision grasp configuration
9. Meijneke C, Kragten GA, Wisse M (2011) Design and performance assessment of an underactuated hand for industrial applications. Mech Sci 2:9–15

Chapter 10
Item Recognition, Learning, and Manipulation in a Warehouse Input Station

Maja Rudinac, Berk Calli, and Pieter Jonker

Abstract One of the challenges of future retail warehouses is automating the order-picking process. To achieve this, items in an order tote must be automatically detected and grasped under various conditions. An inexpensive and flexible solution, presented in this chapter, is using vision systems to locate and identify items to be automatically grasped by a robot system in a bin-picking workstation. Such a vision system requires a single camera to be placed above an order tote, and software to perform the detection, recognition, and manipulation of products using robust image processing and pattern recognition techniques. In order to efficiently and robustly grasp a product by such a robot, both visual and grasping models of each item should be learnt off-line in a product input station. In current warehouse practice, all different types of products entering the warehouse are first measured manually in an input station and stored in the database of the warehouse management system. In this chapter, a method to automate this product input process is proposed: a system for automatic learning, measuring, and storing visual and grasping characteristics of the products is presented.

M. Rudinac (✉) · B. Calli · P. Jonker
Faculty of Mechanical, Maritime and Material Engineering,
Delft University of Technology,
Mekelweg 2, 2628 CD Delft,
The Netherlands
e-mail: m.rudinac@tudelft.nl

B. Calli
e-mail: b.calli@tudelft.nl

P. Jonker
e-mail: p.p.jonker@tudelft.nl

R. Hamberg and J. Verriet (eds.), *Automation in Warehouse Development*,
DOI: 10.1007/978-0-85729-968-0_10, © Springer-Verlag London Limited 2012

Fig. 10.1 *From left to right* Pick-and-place robot, order totes in order picking, zoomed picture of a product tote

10.1 Introduction

Robot arms with end effectors are efficiently used in industry for tasks such as assembling, palletising, and welding. These industrial robots can be characterised by some properties: they are specialised to perform one single task, the conditions in which they operate are strictly controlled, and they work in zones unoccupied by humans [7]. Consider the simple pick-and-place robot of Fig. 10.1: its single repetitive task is to move boxes from one place to another, while it is known in advance which object should be picked and how, and the robot arm is isolated from other robots or humans. One of the challenges of future retail warehouses is robust order picking by robots. In this case, unlike the pick-and-place robots of the metal and electro industry, the objects are not neatly stacked or fixed and also their end pose (3D position and orientation) has some freedom as Fig. 10.1 shows. If it is assumed that various products have to be picked from product totes and be placed in an order tote, these products have to be checked—i.e. is it the correct product?—and their pose has to be determined before they can be grasped by a robot gripper. If a single product type per product tote is assumed, it may occur that the wrong set of products is in the tote, that one or more products are not correct, that a product is damaged, or that unknown objects—not in the product database of the warehouse, such as packaging materials—are mingled between the products. A solution would be to place RFID tags on each item and to use the signal from the tag for localisation of the items in the totes, however, such a solution has several drawbacks. It requires the tagging of all items in the product's production process prior to entering the warehouse, signals from different tags can interfere making it impossible the localise individual items, and such tags still do not provide enough information for grasping, as this depends also on the pose of the item with respect to the other items and the tote.

A more flexible and cheaper solution is to use a camera above the tote and software for localisation and recognition of the items and their most suitable grasping points. Image processing and pattern recognition algorithms can reliably detect partly occluded products, identify those products, and determine their pose under various

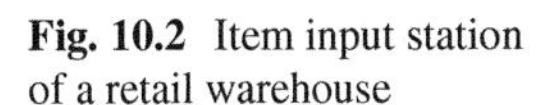

Fig. 10.2 Item input station of a retail warehouse

conditions. In order to do this automatically, it is necessary to learn the characteristics of those products off-line and store the information about their visual appearances and grasping possibilities in a database. This learning process can best be done in a warehouse input station, similar to the one in Fig. 10.2, currently already in use to manually enter new products in the warehouse catalogue.

All new objects go through such an off-line input system in which the properties of the objects, such as weight, volume, softness, etc. are measured and all necessary features for handling the products in the warehouse are stored in that product database. As products can be found in any orientation in the product tote when they need to be picked, the products have to be seen from many orientations in the input system to be of use on-line. Moreover, it is useful for the layout and lighting conditions in the off-line input system to highly resemble the situation in the on-line bin-picking workstation. Therefore it makes sense that the intake of new products in the database is automated with the same robot and vision system that does the actual on-line item picking. However, in contrast to the picking workstation, in the input station a "curious" robot is needed that probes the product such that it can recognise it from various viewpoints and grasp it from various poses. This input system can automatically determine which viewpoints of the products are salient and what are the dominant features that discriminate them from other products stored in the database. For the curious robot in the input station, it is proposed to use a passive vision system with a single camera mounted on a fixed position above the tote, as in the picking workstation. For grasping unknown objects by the curious robot with the goal to change the viewpoint for the top-camera system, visual servoing with an eye-in-hand camera is used until the gripper is collision-free manoeuvred over the object so that the gripper can be closed.

The main challenge in such a curious robot system is the learning and manipulation of objects of which the system has no information; this is still not state of the art. However, there is research on service robots with partial solutions for working in unstructured environments; vision-guided mobile platforms with a robot arm and a gripper for manipulation of objects. The Robotic Busboy [12] is a robot to collect cups from an office table and load them into a dishwasher. A top-view camera above the dishwasher is used for cup detection while a grasp planner picks the cups according to their proximity to the dishwasher. The drawback of this system is that it can only detect cups, and that both cup model and grasping models are calculated off-line. A similar platform is the Stair robot [11]: it can handle more objects and is able to learn object models and grasping points by using synthetic images in a simulated environment.

In this chapter, the design of an off-line curious robot input system for a retail warehouse will be discussed. This system is able to detect, grasp, manipulate, and learn the characteristics of unknown objects in order to grasp them quickly and robustly in an on-line bin-picking workstation.

10.2 System Layout

For the automated input station it is suggested to use a 6-DOF robot arm setup with an underactuated gripper and a single camera placed above a tote with products, as shown in Fig. 10.3. We assume that the illumination condition at the input station is similar or—best—identical to that of the picking workstation. The learning of new products should be performed each time a new product enters the warehouse catalogue; it requires that a single product is placed in a tote to allow the vision system to automatically learn it and store the acquired information in the warehouse management system. Our system is designed such that it can start learning from zero information and is able to slowly fill the database. Its steps are described in the following paragraphs.

Learning starts with localisation of the object. Since the colour of the tote or a table for measuring can vary, first a method is proposed that independently of the environment locates and segments the objects in the scene without having prior information on them or their carrier. As a result, the system draws a bounding box around the object in the camera image and if the camera is calibrated, the bounding box gives an estimate on the object dimensions, which is one of the necessary parameters for the localisation. This is further described in Sect. 10.3.

In a second stage, the segmented objects need to be uniquely described. Since retail warehouse products vary in colour, texture, and shape, e.g. from white boxes till very textured, amorphous shapes, it is very difficult to pick just one method to describe

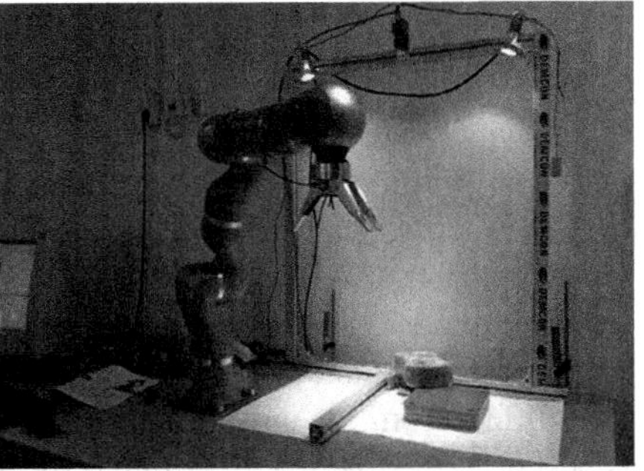

Fig. 10.3 *Left* Robot arm setup at Vanderlande Industries. *Right* Robot arm setup at Demcon

all objects. Consequently, a combination of all possible colour, texture, and shape features is used, and which feature is dominant for which product is automatically calculated. For example, for all white boxes, features that describe high luminance will be dominant. Also a way to calculate the similarity among the many viewpoints of the products is used, which boosts the retrieval of many different products. In a product tote usually many identical objects are placed. To solve the distinction of many *identical* products in different poses, also SIFT keypoints are extracted. The details of the descriptors are found in Sect. 10.4. Finally, since the main purpose of the system is to manipulate the object, also a shape model is calculated that can be used for grasping the item.

Once these ways to describe the object are available, the system should be learnt/trained to recognise it. As the items can appear in the product tote in many different poses, they have to be learnt from many viewpoints. Consequently, the object is learnt while manipulating it. For this the robot should move the object to acquire different viewpoints, while simultaneously the visual descriptors of the product are extracted. The system also needs to learn to classify unknown objects, so when for instance packaging material is present in the tote, the system is able to recognise the outliers. Once sufficient viewpoints of the novel object are extracted, the dominant features are calculated and a database entry is created. This part is described in Sect. 10.5.

Finally, to be able to grasp and manipulate the object, a grasping model must be fit around the unknown object and the best grasping points are calculated. Section 10.6 deals with this.

The final output of the learning stage is a visual description of the object that contains dominant features, extracted keypoints, the size of the object, and a grasping model with best grasping points. This is stored in the database of the warehouse management system. This output is used at a later stage for order picking in the following way: the extracted keypoints are used for object localisation in a cluttered tote, the visual descriptors are used to check that the tote contains the expected products, and the grasping model and grasping points are used to obtain the best strategy to grasp the product.

10.3 Item Localisation

A computationally inexpensive method for localisation of multiple salient objects in a scene without any prior knowledge on their environment [8] is developed. The overall block scheme of this method is displayed in Fig. 10.4, consisting of two stages. In the first stage, a *saliency map* of the scene using a spectral residual method [2] is generated. Salient information means interesting information that pops out from the background, i.e. it differs from the average. An object in a tote is salient as it differs from the uniform colour of the tote. The natural statistics of the scene is best described with a logarithmic spectrum *L(f)* as displayed in the first image of Fig. 10.5.

Now the average information of the scene, *A(f)*, can be calculated using the filters of the second image of the Fig. 10.5, $h^n(1)$, where

$$h^n(f) = \frac{1}{n^2}\begin{bmatrix} 1 & \cdots & 1 \\ \vdots & \ddots & \vdots \\ 1 & \cdots & 1 \end{bmatrix}. \quad (10.1)$$

It is done by convolving the filter with the log spectrum of the scene $A(f) = h^n(f) * L(f)$. Salient information (the spectral residual) is calculated as the difference of the two curves, $R(f) = L(f) - A(f)$ and its result is shown in the third image of the Fig. 10.5. To boost this information, spectral residuals are extracted from the intensity channel, the R-G channel, and the Y-B channel. Using the Inverse Fourier Transform, the output image, called the saliency map, can be constructed in the spatial domain.

In the saliency map of Fig. 10.4c, the bright spots represent interesting points. In order to detect all important information from the saliency map, a Maximally Stable Extremal Regions (MSER) blob detector [5] is applied, which is good in detecting bright regions in dark surroundings. Finally, after detection of the interest points—the yellow points in Fig. 10.4d—the nearby points need to be clustered, as they belong to the same object.

To cluster nearby points, the Gaussian probability kernel is fitted to the image. Since the exact number of objects in the image is not known, the exact width of the probability density function is also unknown. To overcome this a Parzen-window estimation method is used (see Eq. 10.2), where x^i and σ represent the kernel centre and the kernel width respectively, while *N* is the number of detected MSERs contours, and $d = 2$ since every contour centre has two coordinates. Outliers are points with small probability values, which are rejected using Eq. 10.3.

$$p(x) = \frac{1}{N}\sum_{i=1}^{N}\frac{1}{\sqrt{2\pi^d}\sigma^d}\exp\left(-\frac{\|x - x_i\|^2}{2\sigma^2}\right) \quad (10.2)$$

$$\log(p(x)) < \log\left(\frac{1}{N}\sum_{i=1}^{N}p(x_i)\right) - 3\mathrm{var}\left(\log\left(\frac{1}{N}\sum_{i=1}^{N}p(x_i)\right)\right) \quad (10.3)$$

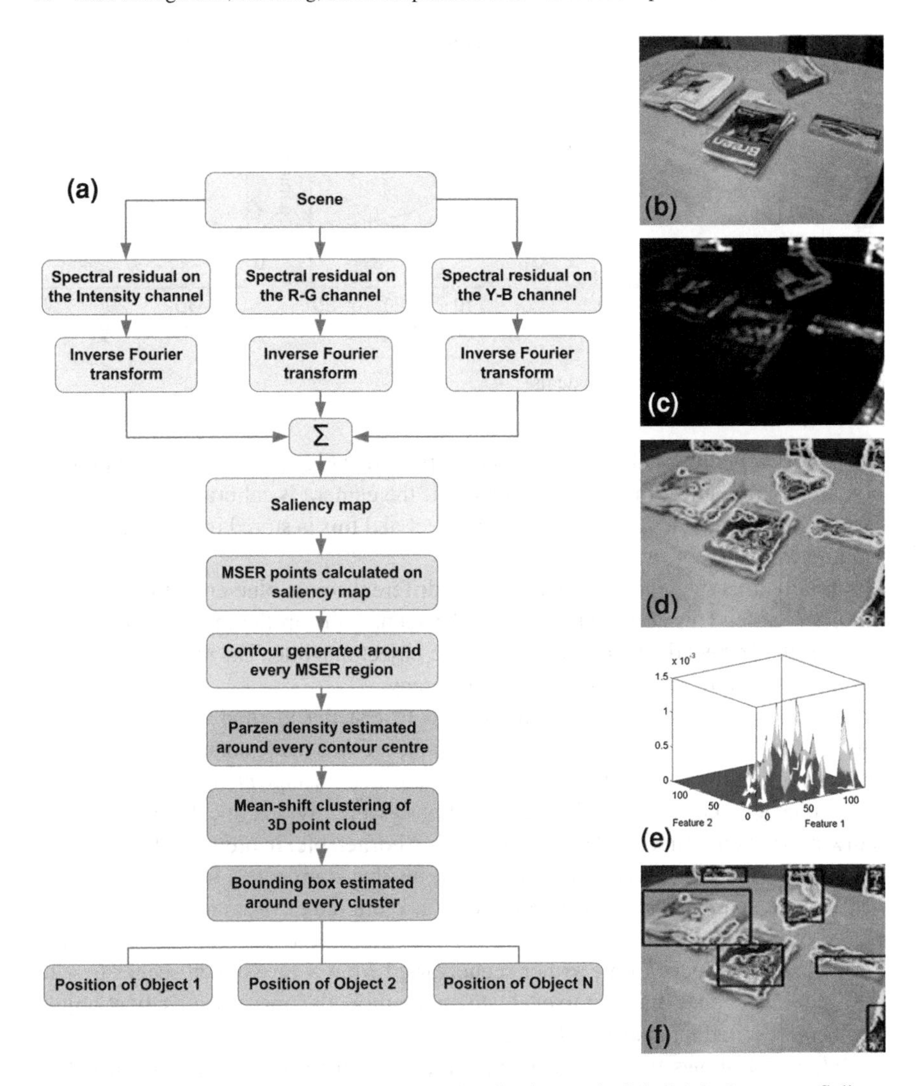

Fig. 10.4 **a** Overall block scheme of the item localisation method, **b** Original scene, **c** Saliency map, **d** Generated interest points, **e** Estimated probabilities, **f** Localised objects

If one observes the 3D plot of Fig. 10.4e, one notices peaks in the diagram in locations that correspond to objects in the scene. However, one also notices small peaks that do not belong to any object. These are outliers and to reject them, criteria are needed to discard low probabilities. All points that are located below the peaks in the 3D plot are grouped, and for this the Mean shift method [8] is used. As a result, several clusters that correspond to objects in the scene are obtained. Finally, the position of the objects is estimated with a bounding box formed by the maximum and minimum values of the pixel coordinates in x and y direction for the contour

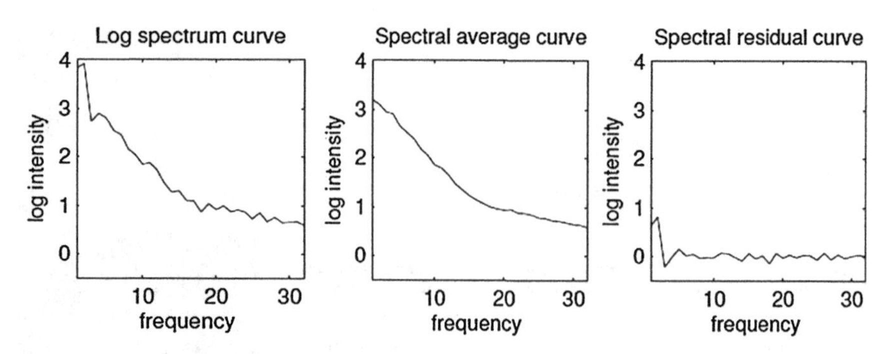

Fig. 10.5 Calculating spectral residuals

centres of every cluster. The final result is displayed in Fig. 10.4f, which is a bounding box around the object in pixel coordinates. If the camera is calibrated, this gives the estimate of the actual dimensions of the object and this is stored in the database as a parameter of the object.

Initially, this method was tested in two different totes (blue and orange) and on an office table. This situation emulated the scenario from the input station, which requires the correct detection of every single object in a tote. The method was found to be very reliable and showed a 100% accuracy.

To make the tests more challenging and to test the method in real-world settings under different lighting conditions, a database was made containing 100 indoor scenes captured by a curious mobile robot that rolled through an office environment, to emulate an AGV in a warehouse. The scenes were divided in categories such as: hallway, kitchen, office, table, entrance, coffee corner, etc. in order to obtain a better understanding of the environmental conditions that dominate a scene. Offices and warehouses share common properties, such as cluttered environments, a large number of objects per scene, illumination changes during the day, etc. (see Fig. 10.6). The average number of objects per scene varied from 8 for kitchen scenes to 4 for coffee corner scenes. In order to evaluate our method, the positions of salient objects had been labelled manually and were used in the tests as the ground truth.

Table 10.1 shows the overall precision and recall results; the precision is high when there is a high number of correctly detected objects and a small number of falsely detected parts of the background; the recall is high when there is a high number of correctly detected objects and a small number of undetected objects. The result is best if the two measures are similar and both high; the F-measure is the ratio between them. Rudinac and Jonker [8] provide more details. Considering that the testing was done in a very challenging environment, a precision between 80 and 90% is a good result. Both a high precision and a very low processing time was found for the "table" and "kitchen" categories, which contained a large number of occluded salient objects. The lowest rates, as well as highest processing times, are observed for; images from the "hallway" category. This is caused by very intensive light from outdoor, which inflicts the saliency detection. This indicates that those images have to be better sieved. The results of the object localisations are presented

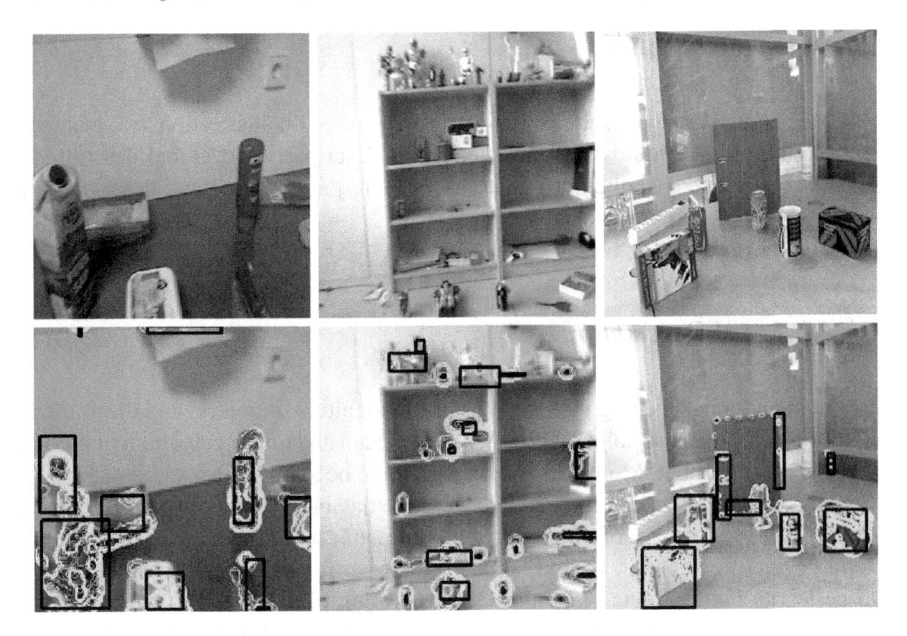

Fig. 10.6 Results of object localisation

Table 10.1 Precision and recall results for different categories of indoor scenes

Category	Precision (%)	Recall (%)	F-measure (%)
Blurred images	90.0	75.0	81.8
Coffee corner	94.3	84.6	89.2
Entrance	80.9	80.9	80.9
Hallway	85.0	87.2	86.1
Kitchen	90.8	83.1	83.1
Office	88.3	85.5	86.9
Table	93.4	92.5	92.3

in Fig. 10.6. Rudinac and Jonker [8] present a detailed comparison of this method with the state of the art, comparing both precision and speed. The conclusion is that this method is very fast compared to the other methods while having similar or better performance. A C++ implementation of the method achieves real-time performance. Tests show that the described saliency detection method cannot only be applied to the localisation and singling out of objects in a tote of an input station, but it can also fruitfully be used to detect and single out incidental salient objects in the case of an AGV dwelling in a warehouse.

10.4 Item Descriptors

After the objects are localised, they need to be described. In this section, two methods to describe visual features are presented (local descriptors and global descriptors) and one method to describe grasp features; they are jointly stored as a descriptor for an object in the warehouse database.

10.4.1 Local Descriptors

In automated order-picking workstations, the recognition of objects and the retrieval of the object's grasping points should be performed in real time, i.e. without hindering the speed of the robot. Moreover, the object should be learnt with a small number of training samples as inserting new objects in the database should take limited time. Finally, a large database of objects should be feasible and maintainable, since retail warehouses often contain around 60,000 or more different items. These items need to be learnt and at a later stage detected from multiple viewpoints, which further increases the processing time. Many descriptors that try to deal with these problems have been proposed in literature. Local detectors proved to be very robust against affine changes and are often used in cluttered real world settings. The Scale Invariant Feature Transform (SIFT) [4] shows the best performance when dealing with objects from different viewpoints, since they offer viewpoint invariance up to a certain extent, such as 40% in change of viewpoint angle. However, the main drawback of local keypoint methods is a large increase in computation time with an increase of the database. SIFT features are very good in object localisation but not for recognition in large datasets. Rudinac et al. [10] designed a method for the efficient reduction and selection of the number of keypoints based on an entropy measure. Although the number of keypoints was reduced by a factor of more than 10, still the matching with large datasets remains a problem. Another reason that using only local keypoints such as SIFT is not enough, is the fact that those keypoints only work with textured objects and cannot be used for the recognition of cases of uniform colour or weakly textured objects. However, due to their value in object localisation, SIFT keypoints are also extracted and stored in the learning stage (Fig. 10.7).

10.4.2 Global Multiview Descriptors

In order to satisfy both the speed and the large database demands, global descriptors were tested, which are fast both in extraction as well as in matching. To make the descriptor fast and discriminative, several global descriptors based on object characteristics such as colour, texture, and edges, are combined [9]. A colour descriptor based on the HSV histogram is used—as it is both robust to size and pose of the object—a grey-level co-occurrence matrix (GLCM) as texture descriptor, and an

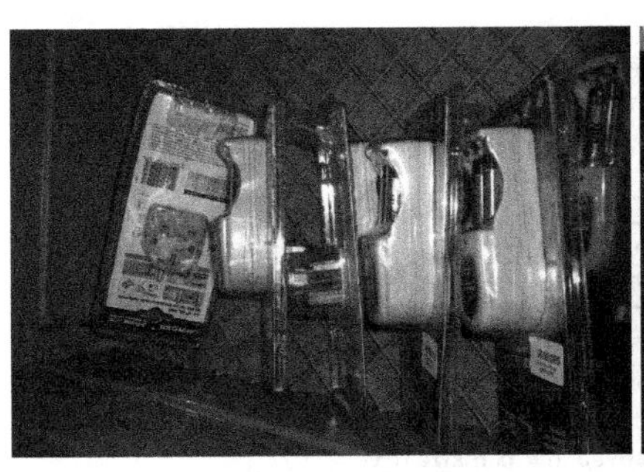

Fig. 10.7 Challenges for object recognition

Table 10.2 Descriptor set

Descriptor	Components	Size
Large descriptor ($L = 252$)	Colour histogram	162
	GLCM	16
	Edge histogram	74
Small descriptor ($L = 25$)	Colour moments	9
	GLCM	16

edge histogram as a shape descriptor. The list of descriptors is displayed in Table 10.2 while a detailed explanation of the specific descriptors is described by Rudinac and Jonker [9].

The final descriptor F_j is a linear combination of the descriptor components from Table 10.2. Furthermore, a normalisation step on all features from the database is performed in order to enhance the influence of the most dominant components and to reduce the influence of the less dominant ones. As displayed in Fig. 10.8, the descriptors from all images from a training dataset are combined in one single matrix $FV(i, j)$, after which that matrix is normalised twice. The first normalisation eliminates the influence of components with large values by dividing each element with the maximum element in the corresponding column, while the second normalisation uses a specific weighing strategy to increase the dominance of features with large variances as is displayed in Eqs. 10.4 and 10.5.

$$W(j) = \frac{1}{\sum_j FV(i, j)} \log_2\left(\text{std}\left(\frac{F_j}{\sum_j FV(i, j)}\right) + 2\right) \tag{10.4}$$

$$FV\text{norm}(i, j) = \sum_{j=1}^{252} FV(i, j) * W(j) \tag{10.5}$$

Detailed formulae are provided by Rudinac and Jonker [9]. These normalisation steps radically increase the precision with almost 30%, especially in situations with noise, occlusion, or when only a small number of training examples is available. Note that an increase in the influence of the most dominant features is very important since in

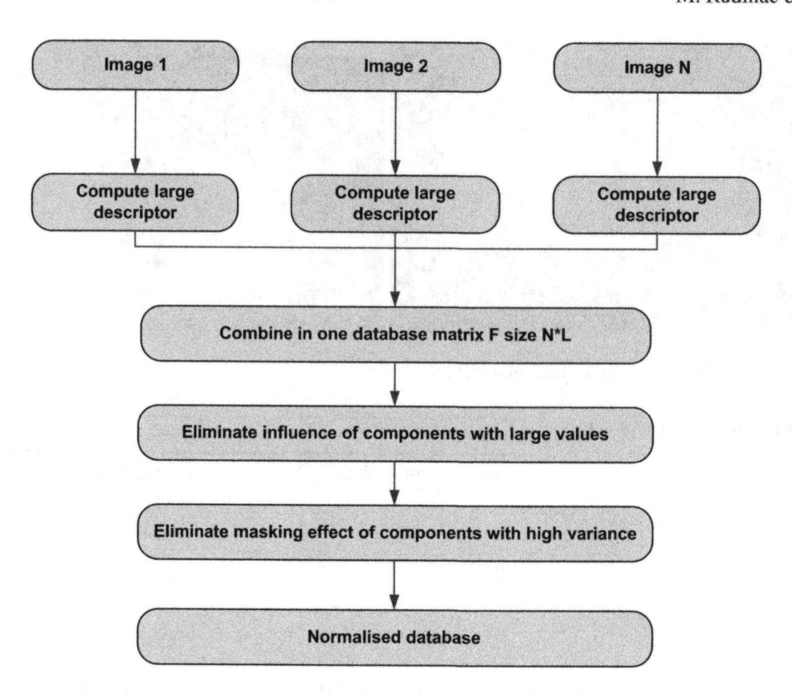

Fig. 10.8 Normalisation procedure

the learning step, the system does not know in advance which features are relevant for a certain product. It can be colour, texture, or shape, or a combination of them, and this method selects them automatically, whereas it also helps in case of noise and illumination changes since the dominant features are very stable.

For testing the Columbia Object Image Library dataset (COIL-100) [6] is used. It consists of 100 objects from multiple viewpoints, whereas the colour images of the objects are taken at pose intervals of five degrees; in total 72 poses per object. This database is very suitable for emulating the bin-picking problem, since it also deals with multiple viewpoints per object. The database was divided in training and test sets, where the normalisation step was performed only on the training set. The descriptor performance was tested on the test set. For all experiments, the results are shown while varying the training set from 75% (54 views per object) to 10% (7 views per object) out of the total database. Also the situation is shown if only 4, 2, or just 1 instance of the object is learnt, which rapidly increases both the learning and matching time. In order to measure the overall recognition performance, all images from the test set are tested; for every image the closest match with the training set is retrieved using the Euclidean distance. When the query image and the closest match found viewpoints of the same object, the match is considered successful. The overall precision is calculated as the sum of all correct matches versus all images in the test set. Table 10.3 shows an extremely high precision for the normalised large descriptor (*top bold row*) and a large raise in precision after using the normalisation step. Also

Table 10.3 Precision for different sizes of training part of the COIL database. The conditions were varied over uniform background, occlusion of 20%, and an illumination change of 50%

Condition	Large vector	Train					Train		
		75%	50%	25%	20%	10%	4 views	2 views	1 view
Uniform	No scaling	99.94	99.94	99.28	98.33	93.70	81.56	55.41	37.61
Uniform	**With scaling**	**99.56**	**99.11**	**98.83**	**98.40**	**97.08**	**92.79**	**84.74**	**76.86**
Occlusion	No scaling	96.22	96.50	92.89	90.65	83.45	72.16	48.93	33.14
Occlusion	**With scaling**	**96.72**	**96.75**	**95.59**	**95.14**	**93.59**	**88.93**	**80.84**	**72.11**
Illumination	No scaling	99.89	99.92	98.94	98.21	93.50	80.15	56.60	41.80
Illumination	**With scaling**	**98.61**	**98.08**	**96.61**	**96.53**	**94.30**	**89.24**	**80.76**	**71.75**

Table 10.4 Speed analysis for the COIL database

Time (ms)	Extraction (per image)	Matching (per image pair)	Matching (with the COIL database)
Large vector	75.0	0.01148	60.1
Small vector	50.8	0.00763	41.3

very promising is the high recognition rate of 93% when training only 4 views of the object.

Further, the descriptor is tested for robustness against problems such as occlusion and illumination change (also see Table 10.3). The table shows that the normalisation gives a higher increase in recognition rate when severe distortions have to be dealt with, indicating that the descriptor is very robust. This is very important for the bin-picking problem, since the objects are learnt (trained) in one setting with specific light conditions and later on picked (tested) in—possibly—deviant settings. Also, as illustrated in Fig. 10.7, occlusions and cluttered environments are constantly present, so being robust for occlusion is highly desirable.

The speed of our descriptor is presented in Table 10.4. It shows that the descriptor is very fast and with an optimised C++ implementation real-time performance even with large databases can be achieved. As is shown in Table 10.4, the matching was performed around 16 frames per second on a database of 7,200 objects. Further tests and a more detailed analysis are described by Rudinac and Jonker [9].

10.4.3 Elliptic Fourier Descriptors

Besides colour, texture, and edge information in order to efficiently describe global shape of objects and determine its grasping model, Fourier Descriptors (FDs) are used. FDs are a widely used modelling technique in pattern recognition to describe the shape of an object. If used together with a similarity transform, FDs are invariant under translation, rotation, scaling, and starting point location. Thus, FD parameters are a good addition to our global descriptors.

Elliptic Fourier descriptors (EFD) provide an easy way of parametrising any closed and bounded curve using FDs and are especially advantageous if a geometric interpretation of the shape is necessary such as when grasping points on objects need to be found [13]. An EFD model of a closed and bounded curve is composed of the sum of sine and cosine functions in different harmonics:

$$x(t) = A_0 + \sum_{n=1}^{k} \left(a_n \cos \frac{2n\pi t}{T} + b_n \sin \frac{2n\pi t}{T} \right), \tag{10.6}$$

$$y(t) = C_0 + \sum_{n=1}^{k} \left(c_n \cos \frac{2n\pi t}{T} + d_n \sin \frac{2n\pi t}{T} \right), \tag{10.7}$$

where a_k b_k c_k, and d_k are Fourier coefficients and k is the number of harmonics.

These coefficients can be obtained using edge data, as explained by Kuhl and Giardina [3]. This EFD model provides the coordinates of the object edge in x and y direction for given t. Also note that these functions are periodic with a period T. The precision of the model is proportional to the number of harmonics which are used to model the object. Figure 10.9 presents a cup and a box modelled using different harmonics. In the learning stage, the Fourier coefficients are recorded and stored with the local and global descriptors. They are used for the object recognition as well, but the parameters are also used for grasping purposes in the bin-picking workstation as explained in Sect. 10.6.

With this descriptor, object classes can be learnt from a small number of examples only, using multiple viewpoints and later recognised in a very fast and robust manner.

10.5 Item Learning

As explained in the previous section, new products must be described from multiple viewpoints in order to be correctly detected in the bin-picking system. In order to automate the learning process, a robot arm navigates towards the localised object in the tote and actively moves the object around to generate the necessary viewpoints for the descriptor. First, after the localisation, the grasp model of the object and its dimensions are extracted and stored in the database. Examples of object localisations are shown in Fig. 10.10. Second, the robot arm is sent in to move the object. The camera above the tote can now segment the object using the motion information obtained from the movement of the object. The background is removed using frame differencing and then form motion history images by collecting motion information from several frames in one single image. However, as the robot arm is moving the object, also the motion of the robot arm is segmented. To filter this out, object tracking is deployed and viewpoint images are made using the overlap of the motion information and the tracking result. If only tracking would be used, parts of the

Fig. 10.9 EFD models of a cup and a box using different harmonics. From *left* to *right* 2, 4, 10, and 20 harmonics are used, respectively. The *green line* denotes the EFD model, the *yellow point* is the model centre (A_0, C_0)

Fig. 10.10 Detected products in a tote

tote could be extracted as the object in situations when the tracker fails. For every viewpoint, all descriptors are extracted and stored.

After acquiring a number of viewpoints—about 10 already yields a good result—normalisation of the database is performed and dominant features are calculated and stored. Also the values of the last performed normalisation are stored. Finally, the object is matched with the database in order to extract statistical parameters to be used for the clustering of objects into object classes, and for calculating if the object is already present in the database. In the order-picking scenario, where objects in the tote are matched with the database to confirm that the correct product shall be picked, the extracted features of the scene are first re-normalised using the stored values on the statistics, and then matched with the database using a nearest-neighbour search to

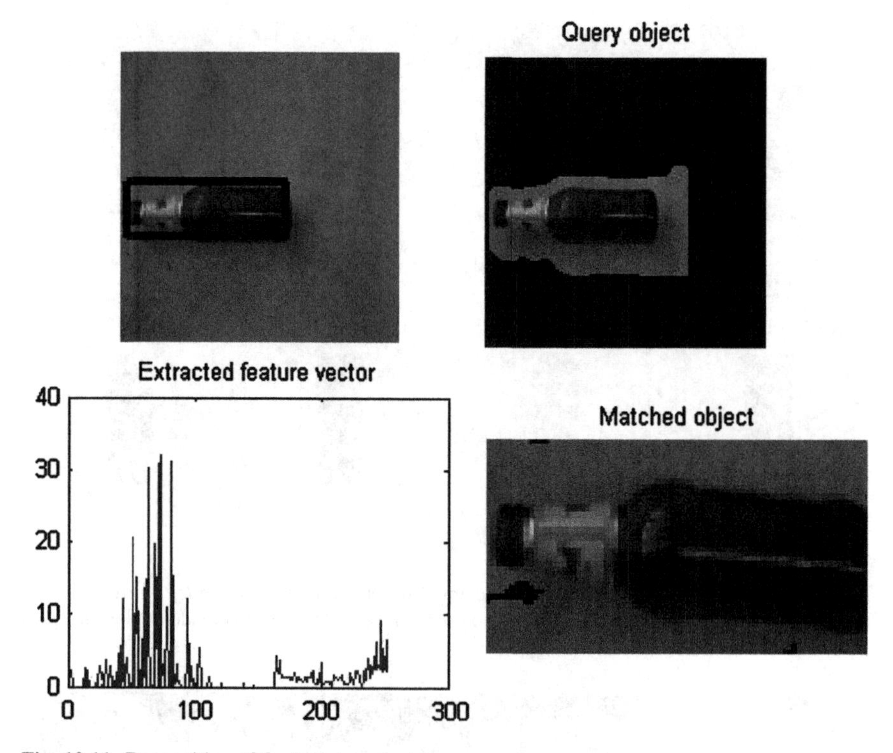

Fig. 10.11 Recognition of the learnt product

find the best matching object. The entire recognition process is shown in Fig. 10.11. First a bottle of beer tracked on a table is shown, then the overlap of tracking and motion mask and query object, then the extracted normalised feature vector, and finally the best matching result from the database is shown.

In order to test whether the active learning scheme works, 20 objects are learnt, while for testing 20 unknown objects are used in addition as distractors. Active recognition means that the recognition of the object is performed using several viewpoints from which each one votes for a best match. In passive learning, the object is learnt using only a single viewpoint acquired from an initial localisation step. For testing retail warehouse objects like the ones presented in Fig. 10.10 are used. The system had to recognise these objects from a database that was actively learnt. Four different scenarios in 100 scenes were tested. From the results displayed in Table 10.5, one can conclude that active learning leads to a large increase in precision and recall results in comparison with passive learning, and therefore active learning is a promising technique for retail warehouse input stations.

Table 10.5 Recognition results on actively learnt data

Learning method	Precision (%)	Recall (%)	F-measure (%)
Active learning—Active recognition	98.0	100.0	98.9
Active learning—Passive recognition	100.0	100.0	100.0
Passive learning—Active recognition	85.7	97.6	91.3
Passive learning—Passive recognition	68.8	94.3	79.5

10.6 Detecting Grasping Points

As explained in Sect. 10.2, with the object descriptor, also Fourier parameters are stored in the object database. In addition to their contribution to the recognition, they are also used to analyse the geometry of the object for grasping purposes. Under the assumption that the concave parts of an object are more suitable for a stable grasp then the convex parts, grasping points can be found by extraction and analysis of the concave points on the object's contour, as explained below.

It is assumed that the robot has an eye-in-hand camera to grasp the object. By way of an elliptic Fourier description of the object's contour, seen from a certain viewpoint, a parametric representation of the curve that forms that contour is available. The first derivative of this function gives the tangent vectors and its second derivative gives the normals in x and y directions. Therefore, for the x direction:

$$T_x(t) = \dot{x}(t) = \sum_{n=1}^{k} \left(-a_n \frac{2n\pi}{T} \sin \frac{2n\pi t}{T} + b_n \frac{2n\pi}{T} \cos \frac{2n\pi t}{T} \right), \qquad (10.8)$$

$$N_x(t) = \dot{T}_x(t) = \sum_{n=1}^{k} \left(-a_n \left(\frac{2n\pi t}{T} \right)^2 \cos \frac{2n\pi t}{T} - b_n \left(\frac{2n\pi t}{T} \right)^2 \sin \frac{2n\pi t}{T} \right), \qquad (10.9)$$

where T_x and N_x are the x-component of the tangent and normal vectors respectively. The curvature of the model can be calculated as the norm of the normal vector:

$$C(t) = |N(t)| = \left| [N_x(t), N_y(t)] \right| . \qquad (10.10)$$

Here the Euclidean norm is used. The local maximum of the curvature data gives the points where the curvature is maximum. However, these points correspond to both the most concave and the most convex regions of the contour. To distinguish the convex and concave regions, the cross product of two consecutive tangent vectors that are selected in the region can be used. The sign of the z-component of the cross product indicates whether the region is convex or concave. If the edge data is read in clock-wise direction, then positive values correspond to concave regions and negative values to convex regions.

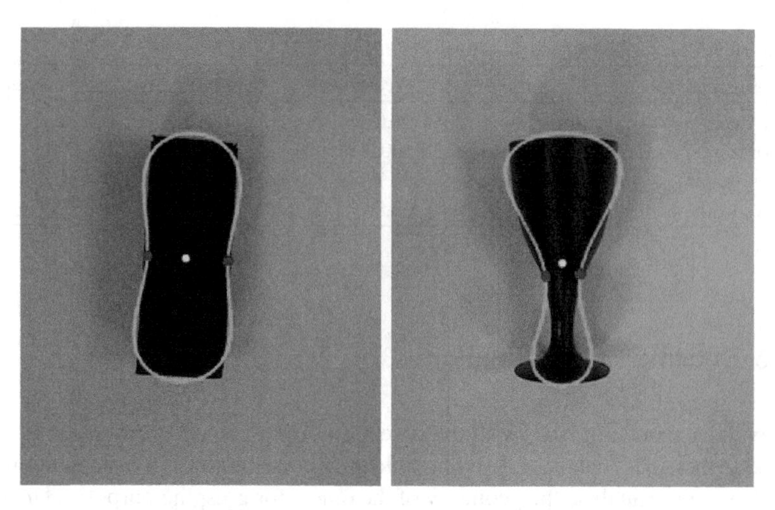

Fig. 10.12 Grasping points detected on a cup and a box. An EFD model with 4 harmonics is used

Points with a first derivative of zero and a negative second derivative correspond to maximum curvature points. Assuming the data is read in a clockwise manner, a set of grasping point candidates **G** can now be obtained by:

$$\mathbf{G} = \left\{C(t) | (C'(t) = 0) \wedge (C''(t) < 0) \wedge ([001](T(t) \times T(t+1)) > 0)\right\}. \tag{10.11}$$

When generating grasping point candidates, using the lower harmonics of the EFD function is preferable. The main reason is that evaluating the global shape of the object is more important for grasping than the details of the curvature. This makes it also possible to find grasping point candidates on flat surfaces of the objects. A function with low harmonics results in concave regions in long and flat surfaces of the objects. This can be seen in Fig. 10.12.

It is important to note that these grasping point candidates do not correspond to 3D points on the contour surface of the object. They are the 2D points of the projection of the object on the camera image seen from a specific point-of-view. However, they are guidelines for the robot fingers. The candidate points can be used for 3D grasping by applying visual servoing to the target. This means that after the initial localisation and recognition of the objects and its approximate viewpoint, the robot can be servoed to the object under the constraint that the object remains approximately in the centre of the image of the eye-in-hand camera and that the curvature of the 2D projection of the object in the image remains the same or is increasing. During the approach of the object, recalculating the EFD parameters will provide updates to the grasping point candidates. If the object covers the whole image, the gripper can be closed [1]. Alternatively, a distance sensor can be attached to the palm of the gripper to detect that the object is close enough to grasp.

10.7 Conclusion

In this chapter, a system for the automatic learning and storing of a descriptor of visual and grasping characteristics of items in a retail warehouse database is introduced. The input station where this learning can be automatically performed consists of a robot arm equipped with a gripper and a single camera placed above a tote. When a new product should be introduced to the warehouse catalogue, it is placed in the tote, after which the product is localised and its size is extracted. To subsequently describe the product for the sake of recognition, a method for object segmentation is derived that requires no prior knowledge on the environment. Since the colour and texture of different products vary a lot, a combination of colour, texture, and shape information is used and it is automatically calculated what the dominant features for every product are. As the object should be recognisable and graspable from many viewpoints, in the learning stage the object is automatically moved by the robot and its features are extracted from multiple viewpoints. For the detection of the candidate grasping points, a shape model of the object is extracted. As database entry information about the size of the product, the features representing its visual appearance, and the most suitable grasping points is presented to the warehouse management system. Tests of the system show promising results and encourage its implementation in future retail warehouse input stations.

References

1. Calli B, Wisse M, Jonker P (2011) Grasping of unknown objects via curvature maximization using active vision, IEEE/RSJ international conference on intelligent robots and systems
2. Hou X, Zhang L (2007) Saliency detection: a spectral residual approach. In: Computer vision and pattern recognition, 2007 CVPR '07 IEEE conference, pp 1–8
3. Kuhl FP, Giardina CR (1982) Elliptic Fourier features of a closed contour. Comput Graph Image Process 18:236–258
4. Lowe DG (2004) Distinctive image features from scale-invariant keypoints. Int J Comp Vis 60:91–110
5. Matas J, Chum O, Urban M, Pajdla T (2004) Robust wide-baseline stereo from maximally stable extremal regions. Image Vis Comput 22:761–767
6. Nene SA, Nayar SK, Murase H (1996) Columbia object image library (COIL-100). Technical Report CUCS-006-96. Department of Computer Science, Columbia University, New York
7. Nof SY (1999) Handbook of industrial robotics. 2nd edn. Wiley, New York
8. Rudinac M, Jonker PP (2010) Saliency detection and object localization in indoor environments. In: Pattern recognition (ICPR), 2010 20th International Conference, pp 404–407
9. Rudinac M, Jonker PP (2010) A fast and robust descriptor for multiple-view object recognition. In: Control automation robotics & vision (ICARCV), 2010 11th International Conference, pp 2166–2171
10. Rudinac M, Lenseigne B, Jonker P (2009) Keypoints extraction and selection for recognition. In: Proceedings of the eleventh IAPR conference on machine vision applications
11. Saxena A, Driemeyer J, Ng AY (2008) Robotic grasping of novel objects using vision. Int J Rob Res 27:157–173

12. Srinivasa S, Ferguson D, Vandeweghe JM, Diankov R, Berenson D, Helfrich C, Strasdat K (2008) The robotic busboy: steps towards developing a mobile robotic home assistant. In: Proceedings of the 10th international conference on intelligent autonomous systems
13. Yip RKK, Tam PKS, Leung DNK (1994) Application of elliptic Fourier descriptors to symmetry detection under parallel projection. IEEE Trans Pattern Anal Mach Intel 16:277–286

Chapter 11
Object Recognition and Localisation for Item Picking

Oytun Akman and Pieter Jonker

Abstract One of the challenges of future retail warehouses is automating the order-picking process. To achieve this, items in an order tote must be automatically detected and grasped under various conditions. A product recognition and localisation system for automated order-picking in retail warehouses was investigated, which is capable of recognising objects that have a descriptor in the warehouse product database containing both 2D and 3D features. The 2D features are derived from normal CMOS camera images and the 3D features from time-of-flight camera images. 2D features perform best when the object is relatively rigid, illuminated uniformly, and has enough texture. They can cope with partial occlusions and are invariant to rotation, translation, scale, and affine transformations up to some level. 3D features can be fruitfully used for the detection and localisation of objects without texture or dominant colour. The 2D system has a performance of 2–3 frames-per-second (fps) at about 400 extracted features, good enough for a pick-and-place robot. Almost all rigid items with enough texture could be recognised. The method can cope with partial occlusions. The 3D system is insensitive to lighting conditions and finds 3D point clouds, from which geometric descriptions of planes and edges are derived as well as their pose in 3D. The 3D system is a welcome addition to the 2D system, mainly for box-shaped objects without much texture or.

O. Akman (✉) · P. Jonker
Faculty of Mechanical, Maritime and Materials Engineering,
Delft University of Technology, Mekelweg 2,
2628 CD Delft, The Netherlands
e-mail: o.akman@tudelft.nl

P. Jonker
e-mail: p.p.jonker@tudelft.nl

R. Hamberg and J. Verriet (eds.), *Automation in Warehouse Development*,
DOI: 10.1007/978-0-85729-968-0_11,

11.1 Introduction

As discussed in Chap. 10, it is assumed for automated order picking in future retail warehouses, that the products have to be checked (i.e. is it the correct product?) and their exact pose has to be determined, before a robot can pick them from a product tote and place them in an order tote. If a single product type per product tote is assumed, it may occur that the wrong set of products is in the tote, that one or more products are not correct, that a product is damaged, or that unknown objects—not in the product database of the warehouse, such as packaging materials—are mingled between the products. In Chap. 10, item learning in an automated input station was discussed; in this chapter, the actual picking in the robot order-picking workstation is discussed. It is assumed that all products having to be picked are already present in the product database of the warehouse management system. This chapter extends the technology of Chap. 10 in the sense that the applicability and robustness of the order-picking system is increased by investigating the combination of 2D and 3D features.

The design challenges for object recognition and localisation for item picking vary along with the classes of objects to detect and the illumination conditions at the order-picking workstation. Local 2D feature-based approaches perform best when the object is relatively rigid, illuminated uniformly without any specular reflections, and has enough texture [7]. They can also cope with partial occlusions and provide invariance against rotation, translation, and scale and affine transformations up to some level. However, they are very sensitive to the lighting conditions, difficult to group, and computationally expensive. 3D features are robust to illumination variance if an active 3D sensor, such as a laser scanner or a time-of-flight (TOF) camera, is used, and can be fruitfully used for the detection and localisation of objects without texture or dominant colour [2]. However, 3D features are sensitive to noise and cannot be used to recognise different objects with similar dimensions, such as two different box-shaped objects with similar width, height, and depth; they will be labelled as the same object. Consequently, combining 2D and 3D features seems to be beneficial for the robustness of the order-picking workstation.

11.2 System Design

Although the challenges and the employed algorithms can vary depending on the object and the environment, the high-level design requirements for a context-free and generic object recognition and localisation system for order-picking can be summarised as follows:

- The algorithms must perform in near real-time; i.e. fast enough to keep the pick-and-place robot busy.
- The system should be equipped with as few sensors as possible to reduce the complexity and the cost.
- The system must localise single objects from a set of multiple objects that are of the same type and in the presence of partial occlusion.

- The accuracy of the recognition and the localisation should be high enough to allow robust order-picking by robots.

To fulfil these requirements, a framework was investigated in which a time-of-flight (TOF) camera was combined with a standard CMOS camera [4]. A time-of-flight camera actively sends modulated infrared light onto the scene and measures the reflections using a high-speed 2D camera. Light from nearby objects arrives earlier than light from objects further away. The output of such a camera is an image in which the intensity of a pixel provides information of the distance of an object point to that pixel. Light pixels are more nearby than dark pixels. Such an image is called *depth image*, *range image*, or *2.5D image*; as one cannot look behind objects the third dimension is only perceived half. Note that due to the noisy signal of the TOF camera, 3D feature extraction requires intense preprocessing and, due to the low resolution of the TOF camera, robust segmentation and recognition using this camera alone remains difficult. Moreover, objects too close to the camera send their light back quicker than the electronics of the camera can handle and the light that is sent out might reflect to some objects in the scene in such a manner that the geometry of those and other objects may become distorted.

The investigated recognition and localisation system is capable of recognising objects that were added to a database in an off-line learning process. In that database now both 2D and 3D features are stored. This learning process is performed in a controlled environment with known background, camera coordinates, and illumination conditions; e.g. in a warehouse input station as discussed in Chap. 10. The decision to use 3D features in addition to 2D features for the verification of the object types is done based on the knowledge of what is supposed to be in the product tote (illustrated in Fig. 11.1). For instance, if objects without texture and dominant colour are stored in the product tote, then 2D features are not applicable and 3D features are used for the detection and localisation of objects. Moreover, if the objects cannot be identified using 2D features alone (due to illumination variance, specular surfaces, etc.) or in case of warehouse management errors such as the wrong items are in the product tote, 3D features can be used in a next attempt.

11.3 2D Image-Based Object Recognition and Localisation

The developed 2D system uses 2D images and various robust (invariant) features to find the objects in a product tote. The building blocks of this system are presented in Fig. 11.2.

In literature many local feature-based recognition systems are described [7]. In the system under study, object recognition is performed on $640x480$ images using scale-invariant features. Both SIFT [6] and SURF [3] features have been tested on their performance in this application to make a decision which was the better candidate. Although SIFT has a slightly better recognition performance than SURF, its high computational load makes SURF a better choice for the core recognition algorithm in the considered system. However, when the recognition fails using SURF, a second

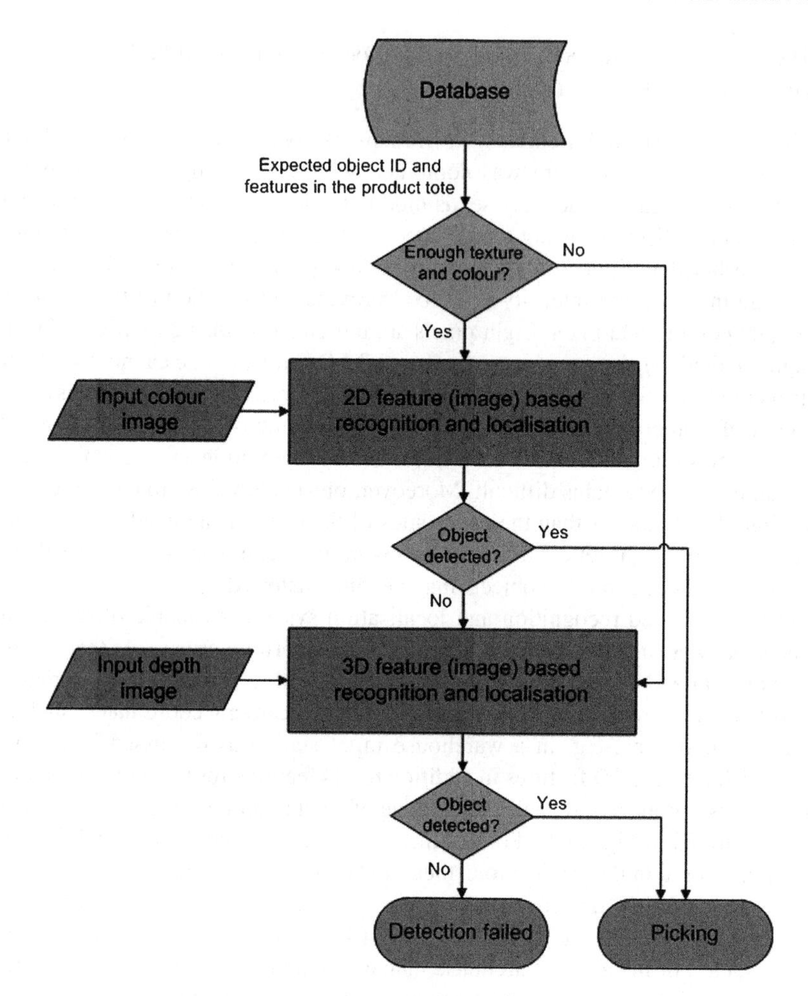

Fig. 11.1 Combined recognition

attempt is done with SIFT to check the tote again. Even more, the performance of the system is further increased by using SURF-128, which has a 128-dimensional descriptor vector. Moreover, the object database is created by using these features extracted from different viewpoints of the object (see also Chap. 10).

For the recognition, 2D local keypoints are found in the input image and corresponding descriptors are extracted. This is followed by keypoint matching, which is done using the method explained in Lowe's seminal work [6]: For each SIFT feature extracted from the input image, the corresponding first and second closest matches in the database are looked up; their ratio gives a measure for the quality of the match. After assigning object classes to each feature via matching, the next step is the clustering of matched features into candidate objects. For fast and effective clustering, a voting scheme was designed. Each feature that was stored in the database in an

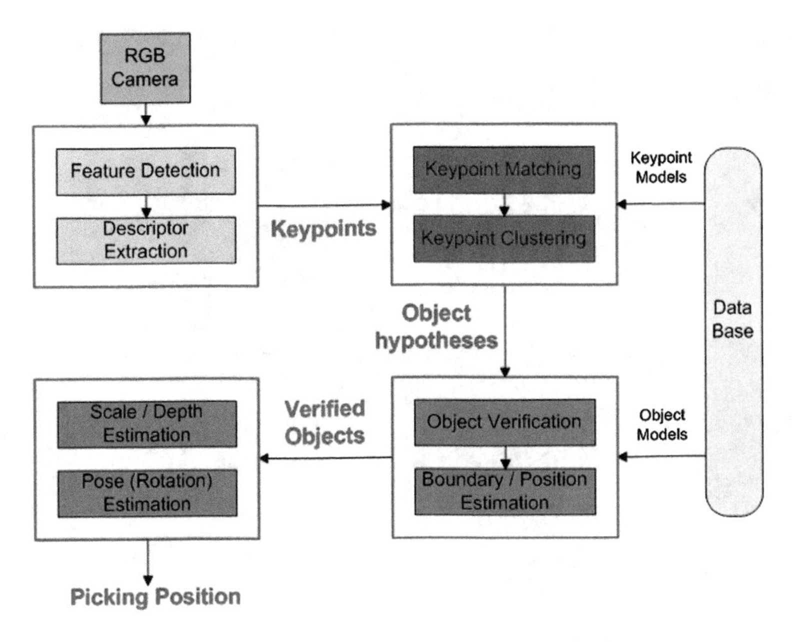

Fig. 11.2 Overall system design for 2D object recognition and localisation

object-learning phase, was stored together with its position with respect to the top-left corner of the object. The input image is divided into bins and each feature votes for a bin (image position) that corresponds to the top-left corner of the object that it belongs to. Subsequently, the bin with the maximum number of votes is selected as the best visible object candidate in the image and the features voted for this bin are labelled as the object features. However, false matches and noise result in model-fitting errors and object loss or false object recognition. Therefore, a random sample consensus (RANSAC) [5] algorithm is used as a verification step to reject outliers (i.e. false matches). Initially, the *homographic mapping* between the object and its model in the database is calculated by using random samples assuming that the object is rigid and has planar surfaces. The homographic mapping is a projective mapping that maps points on one plane to points on another plane and represented by 3×3 matrix, H, which can be derived by using the point-to-point correspondences. The matrix H contains nine entries, but it can be defined only up to a scale. Thus, the total number of degrees of freedom in a 2D projective transformation is eight. Therefore, minimum four points are necessary to compute the H matrix, since a 2D point has two degrees of freedom corresponding to its x and y components. Afterwards, other features are projected onto the model by using the mapping, and the distance between their corresponding matches and their projections is used to evaluate the mapping. The mapping with the minimum sum of distances is selected as the homographic mapping, and points that have large projection distances are discarded. Finally, the boundaries and the pose of each object is calculated using the mapping. The results of the described system are shown in Fig. 11.3.

Fig. 11.3 Recognition based on 2D local features only. *Top left:* 2D feature detection. *Top right:* Best match with object from the database. *Bottom left:* Features voting for the object origin. *Bottom right:* Boundary and object centre determination

This method has near real-time performance, i.e. 2–3 frames-per-second (fps) depending on the number of detected features (e.g. 400). The speed of the algorithm can be increased by using GPU versions of the feature detector. Almost all box-shaped (rigid) items with enough texture can be recognised using local features. The method is also robust against partial occlusions; the boundary and pose of the object can be calculated as long as at least four features are matched correctly. However, this system cannot estimate the pose of non-rigid objects. Although matching can be performed up to a certain level, a homographic mapping and object position cannot be calculated for non-rigid objects. Also, the matching performance decreases with changing illumination conditions since the descriptors of the features are sensitive to lighting.

11.4 2.5D Image-Based Object Recognition and Localisation

In this section, a system is described that utilises 2.5D (range or depth) images to detect objects in cluttered environments in the tote, using a low-resolution time-of-flight (TOF) camera. A TOF camera does not have the time-consuming procedure of matching features from left and right cameras as present in a stereo-camera set-up (see also Chap. 13). Moreover, the built-in infrared light source of the camera

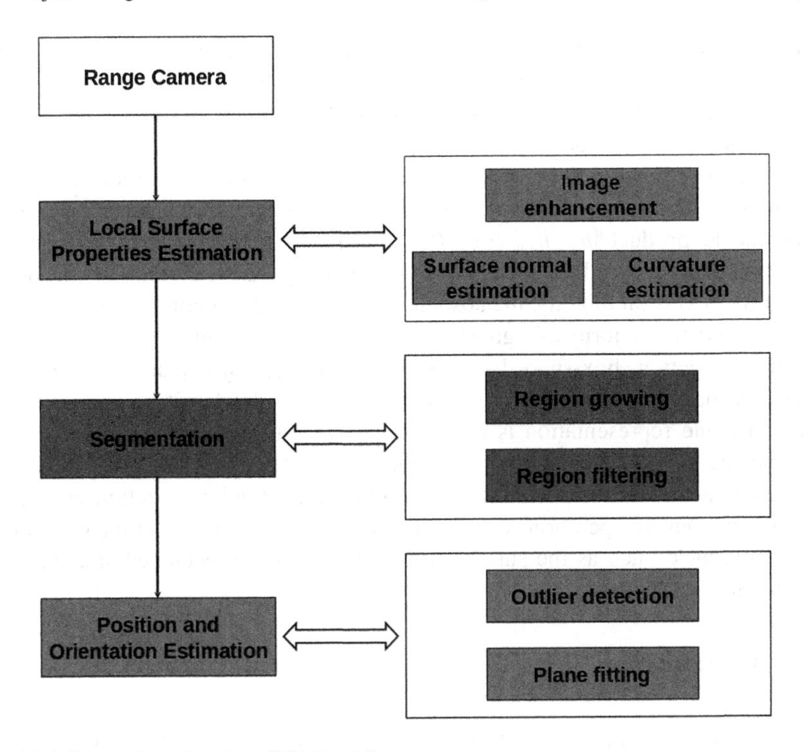

Fig. 11.4 Recognition based on 2.5D local features

makes recognition robust to variations in illumination conditions. It has real-time performance (18 fps), but low resolution (176 × 144), which makes it more appropriate as an additional sensor as opposed to a main sensor. Anyhow, the main benefit of utilising 3D information is that it enables the segmentation and localisation of objects without texture or colour and hence, it eliminates some of the limitations of 2D feature-based methods mentioned above. The building blocks of the developed 2.5D system are shown in Fig. 11.4.

The TOF camera gives point cloud images. These point clouds can be used to retrieve the geometric information that is present in the scene as these clouds represent surfaces [1]. However, the surface properties of the points in the cloud, such as surface normals and curvatures, are defined by their local neighbours rather than defined by a single point. Therefore, the local surface properties must be estimated from the local neighbourhood of a query point. The Eigenanalysis of the covariance matrix of a local neighbourhood can be used to estimate the local surface properties [1].

The local surface properties in a scene can be considered as "similarity measures". Therefore, a region-growing algorithm can be applied that uses the point curvatures, normals, and their residuals, with their user-specified parameters, to group points together that belong to the same surface. It is based on the assumption that points in a linear surface patch should locally make a smooth surface, in which the normal vectors do not vary "significantly" from each other. This constraint can be enforced by having a threshold θ_{th} on the angles between the current seed point and the candidate points that can be added to the region. Additionally, a threshold on residual values r_{th} makes sure that the smooth areas are broken on the edges. A smoothness threshold is defined in terms of the angle between the normals of the current seed and its neighbours. The smoothness angle threshold expressed in radians can be enforced using the dot product $\|n_p \cdot n_s\| \geq \cos(\theta_{\text{th}})$, where n_p and n_s represent the normals of the neighbour point and the current seed, respectively. As the direction of the normal vector has a 180° ambiguity, the absolute value of the dot product is taken.

In the end, the majority of segments grown in the point cloud are planes—assuming that the products are box-shaped—and the centre of each object plane is defined by the centre-of-mass of its constituent points (assuming that the density is homogeneous). Since a plane representation is used, the poses of the segmented objects can be obtained easily. The results of the developed system are given in Fig. 11.5. Different colours indicate different surfaces on which grasping with a suction cup can be performed. During operation, various criteria can be used to select the best surface for a next pick, such as the surface with the biggest non-occluded area and most parallel to the tote floor.

11.5 Conclusions

A product recognition and localisation system for automated order-picking in retail warehouses was investigated, which is capable of recognising objects that have a descriptor in the warehouse product database containing both 2D and 3D features. Combining 2D and 3D features is beneficial for the robustness of the order-picking process. Local 2D features perform best when the object is relatively rigid, illuminated uniformly, and has enough texture. They can cope with partial occlusions and are invariant to rotation, translation, scale, and affine transformations up to some level. 3D features can be fruitfully used for the detection and localisation of objects without texture or dominant colour.

The developed 2D system has a performance of 2–3 fps at about 400 extracted features, good enough for a pick-and-place robot. Almost all box-shaped, rigid items with enough texture can be recognised. The method appears to be robust against partial occlusions; the boundary and pose of the object can be calculated as long as at least four features are matched correctly. The system cannot estimate the pose of non-rigid objects such as objects in plastic bags. Although recognition can be performed up to a certain level, a homographic mapping and object position cannot be calculated correctly. It may be that the information that can be provided is enough for the method for the grasping of unknown objects as described in Chap. 10. However, experiments testing this have not been conducted yet. The 2D matching performance decreases with low illumination conditions as the descriptors of the 2D features are sensitive to lighting. The additional 3D system based on a TOF camera is insensitive to lighting conditions and finds 3D point clouds, from which geometric descriptions of planes and edges are derived as well as their pose in 3D. The centre of each

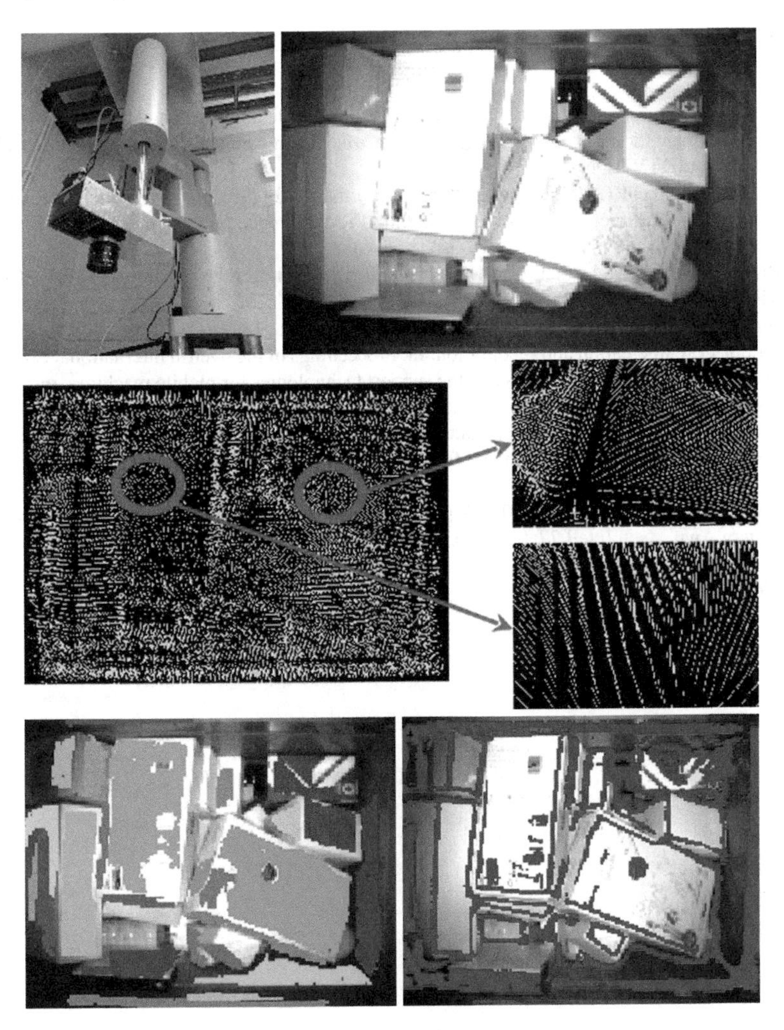

Fig. 11.5 3D object localisation results. *Top left:* 2D camera and TOF camera mounted on a SCARA robot. *Top right:* TOF input image (range image), *Middle left:* Surface normals of the points in the point cloud derived from the range image. *Middle right top:* Low-curvature regions (planes). *Middle right bottom:* High-curvature regions (edges). *Bottom left:* Pseudo-coloured planes. *Bottom right:* Pseudo-coloured edges

object plane is defined by the centre-of-mass of its constituent points, and from this the pose of an object can be easily derived. The 3D system is a welcome addition to the 2D system, mainly for box-shaped objects without much texture or colour. As only laboratory tests have been done at present, more tests need to be done to test the limits of the system in real industrial environments and obtain statistical evidence validating the genuine improvement of the proposed system with regard to state-of-practice systems.

References

1. Akman O, Bayramoglu N, Alatan AA, Jonker P (2010) Utilization of spatial information for point cloud segmentation. In: 3DTV-Conference: the true vision-capture, transmission and display of 3D video (3DTV-CON), pp 1–4
2. Akman O, Jonker P (2009) Exploitation of 3d information for directing visual attention and object recognition. In: Proceedings of the eleventh IAPR conference on machine vision applications, pp 50–53
3. Bay H, Ess A, Tuytelaars T, Van Gool L (2008) Speeded-up robust features (SURF). Comput Vis Image Underst 110:346–359
4. Bayramoglu N, Akman O, Alatan AA, Jonker P (2009) Integration of 2d images and range data for object segmentation and recognition. In: Proceedings of the twelfth international conference on climbing and walking robots and the support technologies for mobile machines, pp 927–933
5. Fischler MA, Bolles RC (1981) Random sample consensus: A paradigm for model fitting with applications to image analysis and automated cartography. Commun ACM 24:381–395
6. Lowe DG (2004) Distinctive image features from scale-invariant keypoints. Int J Comput Vis 60:91–110
7. Mikolajczyk K, Schmid C (2005) A performance evaluation of local descriptors. IEEE Trans Pattern Anal Mach Intell 27:1615–1630

Chapter 12
Integration of an Automated Order-Picking System

Wouter Hakvoort and Jos Ansink

Abstract The large variety of items that is present in a warehouse makes automation of an order-picking workstation a challenging task. It is shown that automation of the item-picking functionality can be achieved by combining an underactuated gripper design and a robust vision system with a commercial-of-the-shelf approach for a robot arm. The demonstrator shows the applicability of the gripping and vision technology developed in the Falcon project. By using weight and stiffness control for the robot and force control for the gripper, the relatively limited accuracy of the vision system can be compensated. The integration of these technologies results in a system that is able to pick a wide range of items, including irregularly-shaped and soft items, and items wrapped in plastic. The realised demonstrator can be used for testing future developments of gripping and vision technologies.

12.1 Introduction

Automation of the order-picking task is a challenging aspect in warehouse automation. The large variety of shapes and textures of different items make the design of a robust and reliable automated item-picking workstation complex.

Item analysis, performed in the Falcon project, indicated that circa 70% of all items are box-shaped. Previous research using an all commercial-of-the-shelf (COTS) approach (see Chap. 8) showed that by using a suction cup, a robot arm, and a vision system, these box-shaped items can reliably be picked. The design of the

W. Hakvoort (✉) · J. Ansink
Demcon Advanced Mechatronics, Zutphenstraat 25,
7575 EJ Oldenzaal, The Netherlands
e-mail: Wouter.Hakvoort@demcon.nl

J. Ansink
e-mail: Jos.Ansink@demcon.nl

R. Hamberg and J. Verriet (eds.), *Automation in Warehouse Development*,
DOI: 10.1007/978-0-85729-968-0_12, © Springer-Verlag London Limited 2012

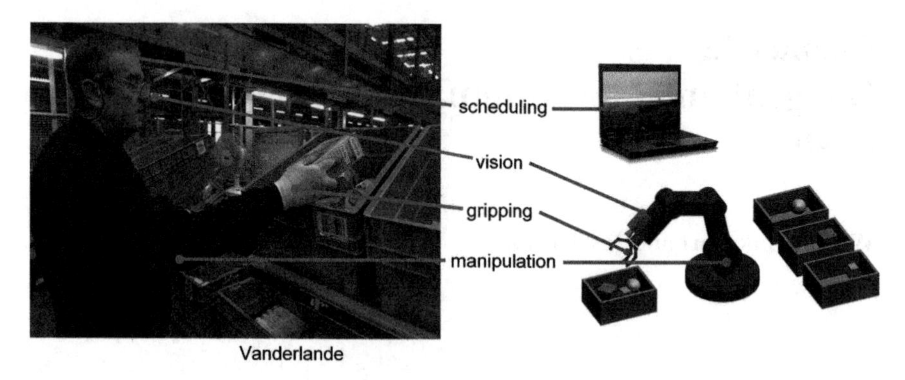

Fig. 12.1 Functions of an order-picking system

order-picking system as described in this chapter will focus on the additional 30% of non-box-shaped items that cannot be handled by the Vanderlande COTS approach.

Development of adequate vision (recognition and localisation) and gripping technologies are identified as the key points to realise a robust and reliable solution for automated order picking of a wide variety of items. A demonstrator setup should show the feasibility of the developments of vision and gripping technology for the order-picking task. Automation of the order-picking task is investigated by combining a mechanical design of an underactuated hand (see Chap. 9), automated texture-based vision (see Chaps. 10 and 11), a COTS approach for manipulation, and adequate system design. This chapter considers the combination of these approaches, the construction of a feasible demonstrator, and its performance.

A schematic overview of the functions involved in order picking is given in Fig. 12.1. Clearly, order picking involves more than grasping and vision. These additional components need to be selected and all systems should be integrated to obtain a working system. This chapter deals with the design and integration of an order-picking setup using the previously mentioned research results. The design and integration has been realised by DEMCON Advanced Mechatronics in close cooperation with the department of BioMechanical Engineering of Delft University of Technology.

12.2 System Engineering

The engineering and construction of the order-picking system is preceded by a system engineering step. The system engineering involves the formulation of the system requirements, the conceptual solution for the realisation of these requirements, the selection of the components, and the definition of the interfaces between the components and their surroundings.

Fig. 12.2 Items to be picked

12.2.1 System Requirements

The formulation of the system requirements is based on the function of the system and the properties of the items to be picked. The Vanderlande demonstrator setup showed that rigid and box-shaped items can be picked well using a suction cup (see Chap. 8). Box-shaped items cover only part of the items handled in a warehouse. The targeted order picker should handle a wider range of items, including irregularly-shaped items, soft items, and items wrapped in plastic. On the other hand, it is assumed that the items are non-magnetic, dry, clean, and not sticky or fragile. Furthermore, the weight and size are limited to 0–500 g and 55–130 mm, respectively. Examples of typical items to be picked are shown in Fig. 12.2.

Besides the definition of the items, the specification of the system is formulated using a functional decomposition of the order-picking task. The functional decomposition is used to formulate a set of use cases with increasing complexity. These use cases are used to structure the development process of the order-picking demonstrator, which manages the risks in the development process by gradually building the system with intermediate evaluation moments. The defined use cases are:

A Grasp a product from a predefined location.
B Grasp a product localised by vision.
C Grasp the best oriented product localised by vision.
D Rearrange the products if no product is accessible.

The first two use cases involve realisation of hardware, the last two use cases involve extension of software.

Besides the functional requirements, the interface of the system with its surroundings is defined, e.g. the robot should not injure a person and the system should have an external interface with either a user or a central warehouse system.

12.3 Conceptual Design

A gripper and a vision system were specifically designed for the considered order-picking problem (see Chaps. 9, 10 and 11, respectively). Due to time restrictions, the gripper presented in Chap. 9 was not implemented in the demonstrator. Instead, an earlier version of this underactuated gripper was used, which is less specific for this application.

Besides these subsystems, components for manipulation and scheduling are required. Considering the specific requirements on those functions imposed by the application, suitable off-the-shelf components were selected. All components are displayed in Fig. 12.3 and discussed hereafter.

Vision System The vision system (see Chaps. 9, 10, 11) consists of a commercial 2D camera (Fig. 12.3a) and a software library. The software makes use of SIFT [5] or SURF keypoints [1]. These keypoints define scale-invariant properties of the texture of items. An item is detected by finding keypoints in the image that match distinctive keypoints of an item stored in a database and clustering the keypoints of a single item using Hough clustering. After recognition, the location of the item is determined from the orientation and the relative distance of the keypoints. The algorithm is intrinsically insensitive to displacements and rotations in the plane of the camera and is accurate in these directions. Out-of-plane rotations of the item should be limited (maximum 40°.) and the detection of the distance to the item and the out-of-plane rotations is relatively inaccurate.

Gripper The gripper (Fig. 12.3b) consists of three fingers each with two phalanges (cf. Chap. 9). Only one motor is used to actuate the gripper. This underactuated design distributes the motor force over the six phalanges. The gripper is constructed such that it can grasp an item using a power grasp. Setting the motor current directly defines the pressure exerted on the item. The gripper is simple and robust, yet it realises a stable grasp for a wide range of items with diameters between 55 and 130 mm and masses up to 2 kg.

Robot Arm The KUKA Lightweight robot (LWR) arm [2] (Fig. 12.3c) was selected for the automated order-picking system. This robot combines several properties that make it particularly suited for the considered application. The robot is light-weight, which enhances safety, reduces power consumption, and enables future mobile application. The human-inspired design with seven degrees of freedom results in the ability to reach around obstacles. Each of the robot's joints is equipped with torque sensors. These sensors enhance safety by their use for collision detection and compliance control. Compliance control means that the robot can be programmed to behave like a spring. This allows the robot to operate in an ill-defined and unstructured environment. Furthermore, the sensors can be used to verify the grasping of the correct item by determining its weight. Finally, KUKA provides various well-defined ways for interfacing via Ethernet, which allows easy integration with a GUI or warehouse control system.

Industrial PC An industrial PC (Fig. 12.3d) was used to implement the vision algorithm and the GUI, to interface with the robot and to do the scheduling of the

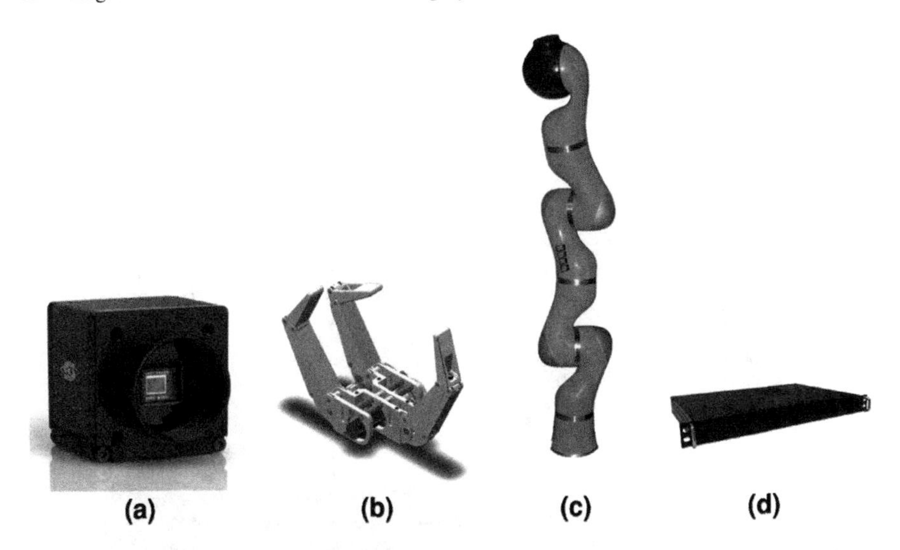

Fig. 12.3 Components of the order-picking demonstrator. **a** Camera of the vision system. **b** Gripper. **c** KUKA Lightweight robot arm. **d** Industrial PC

order-picking tasks. Furthermore, OpenRAVE [4] was implemented on the system. OpenRAVE is an open-source software platform for path planning, collision detection, visualisation, and simulation of robotic systems.

To facilitate integration of these four components, several interfaces were defined. This is a critical part of the system integration process. The interfaces are used as guidelines for hardware, software, and electronics design. In this way, modules can be designed separately in each domain without compromising the system integration at a later stage in the project. Figure 12.4 represents a schematic system overview with the different interfaces specified (M = Mechanical, E = Electrical, S = Software). The different coloured blocks are: base (blue), manipulator (red), low-level control (orange), high-level control (light blue), user interface (yellow), mobile controller (pink). In the realised system, a static base was used instead of a mobile base to reduce development time. This is indicated by the pink colour of the mobile controller block.

12.4 Implementation

This section describes the implementation that realises the conceptual design presented in the previous section.

12.4.1 Hardware Engineering

Based on the defined specifications and interfaces, a mechanical and electrical design was made for the demonstrator platform (Fig. 12.5). The mechanical design consists

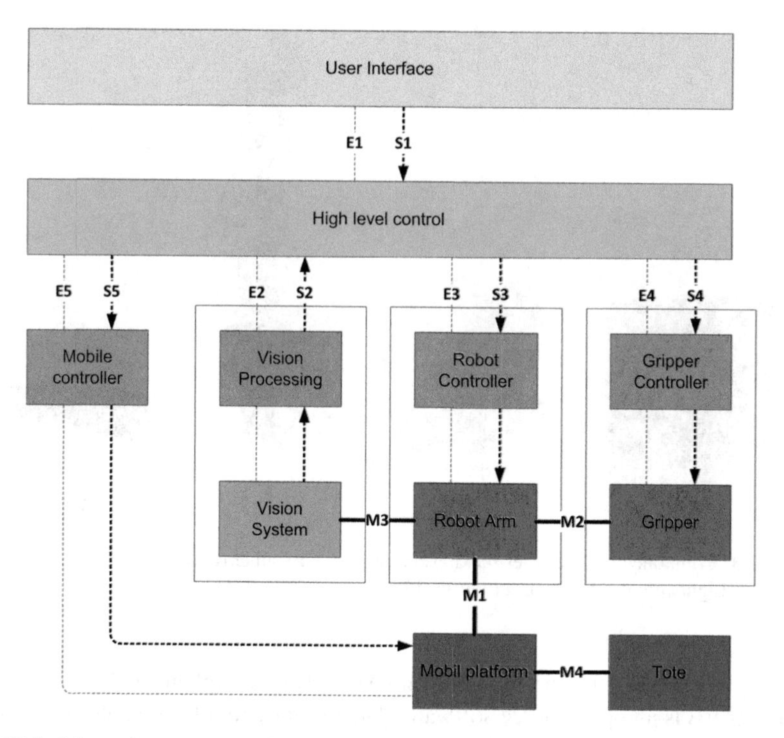

Fig. 12.4 Schematic system overview with the various interfaces

of several interconnected modules that can be easily assembled and disassembled for transportation. An electronics cabinet is located in a small cart that is an integral part of demonstrator mechanics. Polycarbonate windows are placed around the robot with additional safety switches in the doors and sliding windows to guarantee human safety during autonomous movement of the robot system without limiting visibility at a fair. Dedicated lighting is mounted inside the polycarbonate box to make the vision system robust with respect to reflections and ambient light. Finally, the electronics cabinet contains the power distribution, safety electronics, an industrial PC, the KUKA controller, and an output module and amplifier to operate the gripper.

12.4.2 Software Engineering

The software architecture of the order-picking demonstrator is represented in Fig. 12.6a. The central module in this design is the main controller manager. This main controller handles the different high-level control tasks like task scheduling and system monitoring. Three software modules are added to support the main controller and to supply interfaces to the vision system, the robot controller, and the

Fig. 12.5 Mechanical design

path-planning software. These interfaces were created using industry standards and open-source software like TCP, XML, OpenRAVE [4], and OpenCV [3]. The system can be operated via a GUI. This GUI is separated from the main controller and connected via a network protocol. This way the GUI can easily be replaced by, e.g. the high-level control system of a warehouse.

Several modifications were made to the vision software for integration in the order-picking system. First of all, the software was reorganised to allow integration with the main controller software. Restructuring of the software resulted in faster execution times. Moreover, the software was improved with a better clustering algorithm and the ability to recognise multiple (overlapping) items of the same kind. Furthermore, the database containing the item's features was extended with grasping features. These features consist of the location to grasp the product and the gripper's motor current, which defines the force exerted by the gripper on the item.

The KUKA robot controller takes care of the real-time control of the robot. The robot controller is programmed to perform a certain set of tasks, e.g. move to a specified point, wait a few seconds, and close the gripper. These tasks can be activated from the main controller manager over a network interface. To create a robust and reproducible order-picking application, two additional features were implemented on the controller. Firstly, the compliance control mode of the KUKA LWR is used to cope

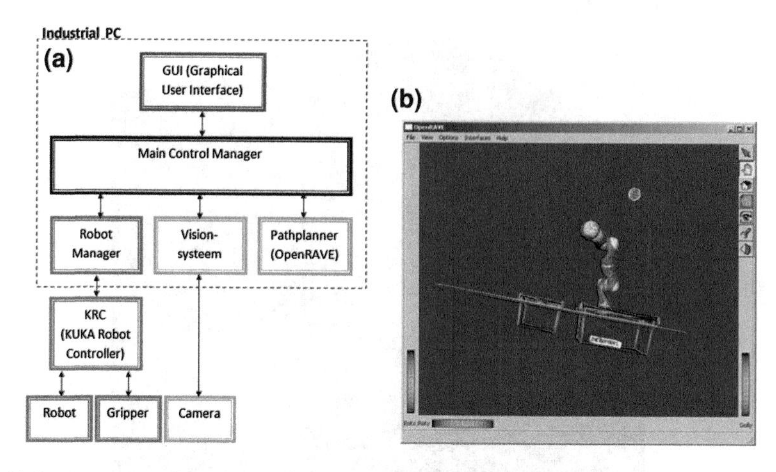

Fig. 12.6 Elements used in software engineering. **a** Schematic overview software architecture. **b** OpenRAVE environment

with the inaccurate depth estimation of the vision system. Due to this compliance control mode, the robot exerts only a limited force when it is not able to reach the specified location due to a collision. Secondly, the robot can be commanded to weigh the item held by the gripper. This feature is used by the main controller to check whether only a single product is held by the gripper.

Currently, OpenRAVE is programmed to visualise the robot moves on a monitor (Fig. 12.6b). In the future, the OpenRAVE environment can be used to plan the motion of the robot.

The main controller manager schedules the picking strategy of the order-picking demonstrator. In short, the picking strategy can be described as follows: The vision system is used to determine the position and orientation of an item. After detecting the item, the main controller will instruct the robot manager to grasp the item using the robot and gripper. Currently, predefined robot movements are used to grasp items.

12.5 System Test Results

The realised order-picking demonstrator is depicted in Fig. 12.7. The system is able to perform its task: The presence of a requested item in the product tote is verified, its location is identified, the gripper is manipulated towards the item, the gripper is closed, the item is manipulated upwards, the item is weighed, the item is manipulated towards a position above the order tote, the gripper is opened, and the manipulator moves backs to its initial position, such that the sequence can be started again (Fig. 12.8).

The whole cycle takes 20 s, where about 2 s is required for (vision) processing. Most time is required for the movement of the robot arm. Faster motions were tested, but then occasionally speed limits for the robot joints were exceeded. A higher

Fig. 12.7 Falcon demonstrator

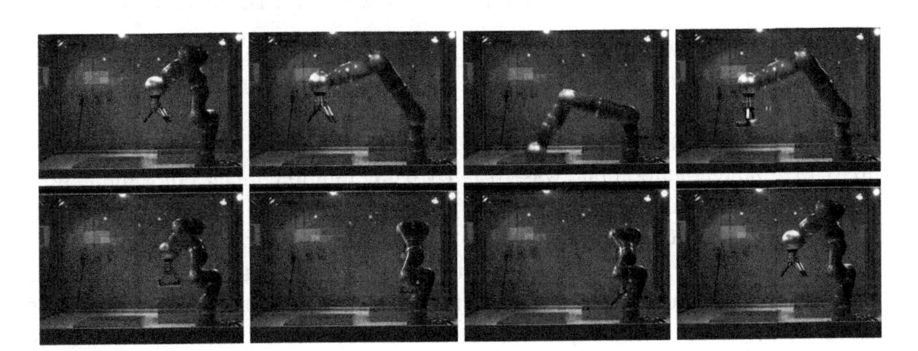

Fig. 12.8 Motion capture of a pick-and-place cycle

velocity and more efficient robot motions (shorter distances) require more extensive path planning, using for instance OpenRAVE.

The designed vision algorithm is independent of 3D location and planar rotations. The most important requirement for successful item recognition using this vision algorithm is a sufficient number of features present on the item to be recognised. With this requirement fulfilled, the vision has a recognition rate of 90% for items

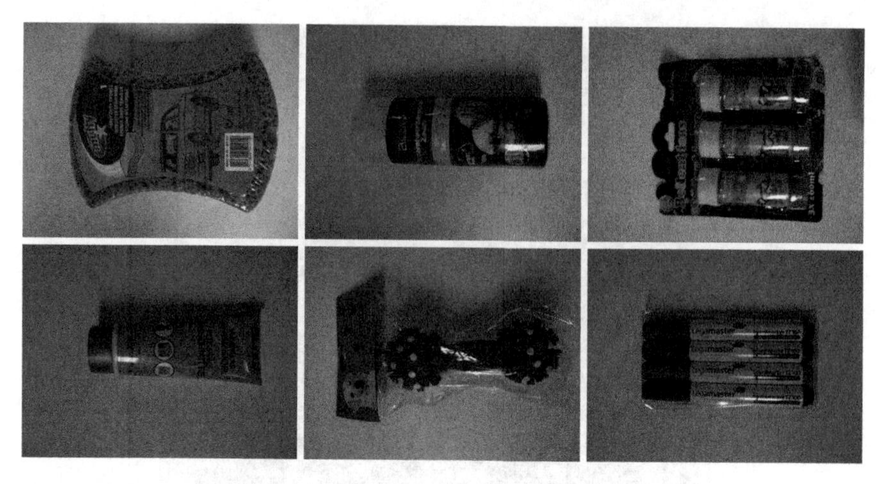

Fig. 12.9 Items that could be picked

Fig. 12.10 Items that could not be picked

that have a non-planar alignment of up to 30°. For items with a non-planar alignment up to 45°, the recognition rate was 70%. By using multiple images of a single item under different angles, a recognition rate of 100% is achievable. Additionally, items may overlap up to 50% without degrading the recognition performance.

The used gripper is able to grasp all items that are within the grasp limits of the gripper. It is also able to do reliable precision grasps with several items after applying rubber on the finger tips to improve the friction coefficient. Figure 12.9 shows some examples of items that can reliably be picked.

The system was able to grasp approximately half of a set of predefined products that were challenging for picking because of their irregular shape, softness, and/or plastic wraps. The items that could not be grasped had no distinctive texture for the vision system (e.g. uniformly-coloured items), or they were outside of the grasp limits of the gripper. Figure 12.10 shows a few items that could not be picked.

The system did not pick up a wrong item, but occasionally an additional item or no item was picked. These cases can be identified using the weighing functionality of the robot. Sometimes the procedure was cancelled, because the manipulator was commanded to move outside its working range, and once it hit the tote with one of the gripper fingers. More extensive path planning is required to prevent these malfunctions. The use of stiffness control for the robot and force control for the gripper proved to be a good combination to cope with the relatively limited accuracy of the vision system.

12.6 Discussion

A demonstrator for automated order picking has been successfully realised. The demonstrator shows the applicability of the gripping and vision technology developed in the Falcon project (see Chaps. 9, 10, and 11). It has been shown that the integration of these technologies and a suitable manipulator results in a system that is able to pick a wide range of items, including irregularly-shaped and soft items, as well as items wrapped in plastic. Essential for a timely realisation of the demonstrator was multi-disciplinary system engineering, which involved the definition of the overall system requirements, the subsystem requirements, and their interfaces on electrical, mechanical, software, and optical level, stating the desired system functionality.

The demonstrator can be used for testing future developments of gripping and vision technologies. The performance of the demonstrator would largely benefit from further development of its path-planning functionality, because this improves the robustness and speed of manipulation. The demonstrator should not be used in a warehouse system in its current realisation. The selected components and the software implementation should be revised on aspects as robustness, lifetime, and ability for integration with the hardware, software, and logistics of the warehouse. For example, it may be necessary to extend the order-picking system with a movable base to be able to integrate it with the current state-of-the-art warehouse architectures. Integration of the newly designed underactuated gripper will further improve robustness and reliability (see Chap. 9).

References

1. Bay H, Ess A, Tuytelaars T, Van Gool L (2008) Speeded-up robust features (SURF). Comp Vis Image Underst 110:346–359
2. Bischoff R, Kurth J, Schreiber G, Koeppe R, Albu-Schäffer A, Beyer A, Eiberger O, Haddadin S, Stemmer A, Grunwald G, Hirzinger G, (2010) The KUKA-DLR lightweight robot arm—a new reference platform for robotics research and manufacturing. In: Proceedings for the joint conference of ISR 2010 (41st International Symposium on Robotics) und ROBOTIK 2010 (6th German Conference on Robotics)
3. Bradski G, Kaehler A, (2008) Learning OpenCV: Computer Vision with the OpenCV Library. O'Reilly Media, Inc, Sebastopol
4. Diankov R, (2010) Automated construction of robotic manipulation programs. Ph.D. thesis, Carnegie Mellon University, Robotics Institute, Pittsburgh
5. Lowe DG (2004) Distinctive image features from scale-invariant keypoints. Int J Comp Vis 60:91–110

Part V
Transport by Roaming Vehicles

Chapter 13
Self-localisation and Map Building for Collision-Free Robot Motion

Oytun Akman and Pieter Jonker

Abstract Robotic manipulation for order picking is one of the big challenges for future warehouses. In every phase of this picking process (object detection and recognition, object grasping, object transport, and object deposition), avoiding collisions is crucial for successful operation. This is valid for different warehouse designs, in which robot arms and autonomous vehicles need to know their 3D pose (position and orientation) in their environment to perform their tasks, using collision-free path planning and visual servoing. On-line 3D map generation of this immediate environment makes it possible to adapt a standard static map to dynamic environments. In this chapter, a novel framework for pose tracking and map building for collision-free robot and autonomous vehicle motion in context-free environments is presented. First the system requirements and related work are presented, whereafter a description of the developed system and its experimental results follow. In the final section, conclusions are drawn and future research directions are discussed.

13.1 Introduction

One of the challenges of future warehouses is robust order picking by robots (see Chap. 8). Four phases can be distinguished in this task: collision-free object detection and recognition, collision-free object grasping, collision-free object transport, and collision-free object deposition. These phases are applicable for the two flavours of warehouse design, i.e. man-to-goods and goods-to-man types of systems.

O. Akman (✉) · P. Jonker
Faculty of Mechanical, Maritime and Materials Engineering,
Delft University of Technology, Mekelweg 2, 2628 CD Delft, The Netherlands
e-mail: o.akman@tudelft.nl

P. Jonker
e-mail: p.p.jonker@tudelft.nl

R. Hamberg and J. Verriet (eds.), *Automation in Warehouse Development*,
DOI: 10.1007/978-0-85729-968-0_13,

In a man-to-goods system, a robot on an automatic transport system may travel along a path to come close to the object, grasp it, transport it elsewhere, and deposit it in some output bin. In goods-to-man systems, the robot and transport systems are separated; the collision-free object transport phase is now performed by totes on conveyor belts. For moving objects —either in the gripper of a robot manipulator, or in a tote on an Automated Guided Vehicle (AGV)—both transport systems—the robot and the AGV—need to be aware of the 3D map of the scene—the static and moving objects in the scene as well as their own positions in that scene—in order to properly plan collision-free paths. By way of example, the robot can collide with itself, crash into the wall of the tote or into objects in input or output totes. If humans are in the neighbourhood also they have to be avoided. Similarly, AGVs might collide with walls, furniture, other AGVs, or humans. Note that for collision-free path planning it is not necessary to recognise the objects; only the geometry of the scene needs to be found in order to manoeuvre within it in a collision-free way.

In this chapter, a system is described that is able to track the 3D pose of a moving stereo-camera pair in a 3D world, while simultaneously building a 3D map of that world. The stereo pair can be mounted on AGVs as well as on robot arms for collision-free motion and path planning in both man-to-goods and goods-to-man warehouse solutions. We are well aware of the fact that also pre-programmed maps can be used as basis for path planning. However, similar to the reasoning of object recognition versus object verification, the fact that knowledge as well as sensing is used instead of knowledge only, makes the system more robust. When the map is incidentally incorrect—a fixed object has moved, a tote is not in its expected position—the system can recover from this, proceed as well as it can and signal the error. Moreover, the system is able to detect and cope with changes in the environment, such as moving humans and vehicles. Also in this case, foreknowledge can be used to speed up the search for an optimal path as this foreknowledge indicates the most likely situation. Consequently, this chapter studies context-free 3D pose tracking and map building for collision-free robot motion.

First, the requirements of pose tracking and map building for different scales and situations in warehouses are specified (Sect. 13.2). Second, these requirements are applied to algorithm development and a generic design for 3D pose tracking and map building is presented (Sect. 13.3). Finally, the results of the system tests and an overview of future work are presented (Sect. 13.4).

13.2 Design Requirements

The design of a 3D pose estimation and mapping system is highly dependent on the scale of the operation. In warehousing, this scale typically ranges from visual servoing for in-tote picking by a robot arm at a small scale to path planning for transport by AGVs at a large scale.

The design challenges for those different scales differ due to hardware and environmental constraints. For instance, the complexity and computational load of the system increases as the area in which the robotic agent can operate increases and hence the accumulated pose estimation error becomes an issue; i.e. the pose error with respect to the start pose builds up. Also the performance of the depth estimation using a stereo setup decreases as the distance between the cameras and the scene increases. Depth estimation with a pair of cameras is performed by matching objects, keypoints or pixels in the left image by the identical objects, keypoints or pixels in the right image and then performing triangulation on the differences (the *disparity image*) to obtain the distance (or depth) of objects, keypoints or pixels with respect to the cameras.

For this the knowledge of the distance between the cameras is used. The depth range and accuracy is a function of the distance between the cameras. The processing time and depth precision is a function of the size of the pixel matrices of the cameras. For AGVs roaming around in a factory setting, a large depth range is necessary and hence a large distance between the cameras. Visual servoing of a robot arm towards an object for the sake of grasping, takes place in a relatively small area and hence the cameras can be close together. The large movements in comparison with the operational area of the arm introduce large pose estimation errors. Also a small number of available *landmarks*—remarkable points that can be used for stereo matching—i.e. their sparseness in a small area such as a tote, decreases the accuracy of both pose estimation and mapping. Although the challenges and employed algorithms can vary depending on the scale of the operation, the high-level design requirements for a context-free and generic system can be summarised as follows:

- The algorithms must perform in real-time or near real-time.
- The system should be equipped with as few sensors as possible to reduce the complexity and the costs.
- The accuracy of the pose estimation should be sufficient for a mobile platform to accomplish its task.
- The accuracy of the constructed 3D maps should be sufficient for the task at hand and they should be self-adjusting to the environment.

In dynamic scenes, it is highly desirable for a robotic agent to understand its environment and act as quickly as possible. Given the current state-of-affairs of robotic movements, a reasonable and clear performance goal is to get as close as possible to the speed of humans to achieve a similar throughput (e.g. pick-and-place operations per hour). Also, the required accuracy for both pose estimation and mapping is highly dependent on the task and therefore should be self-adapting. For instance, a pose estimation in millimetres resolution may not be required for an AGV travelling in the warehouse. However, such a precision is important when a grasp action is considered. Finally, the level of detail of the created 3D maps also depends on the environment and the task. For example, high accuracy and density are not required when approximately planar surfaces such as walls are reconstructed, whereas more detailed information is necessary when highly unstructured surfaces or environments should be modelled. Moreover, it will be clear that a map, representing the interior of a tote for collision-free object grasping with a robot hand, is of a different scale than an environmental map for collision-free motion of an AGV.

The topic 3D pose estimation from fiducials (i.e. markers) has been extensively researched in the past [5]. With these markers and a single camera, the pose of that camera can be determined up to a scale factor. The marker detection is based on the search of contours of (possibly skewed) rectangular objects in the camera, followed by the detection of a valid pattern (ID) in its contour's interior. The four corner points of a marker are used to calculate the rotations of the marker with respect to the camera. In order to obtain a higher accuracy of the positions of the corners than the size of a single pixel, *sub-pixel accuracy*, the corners of the skewed rectangle are omitted, whereupon high accuracy line fits—using a Gaussian line model on the grey-value pixels that form the lines—are done on the four line-pieces making up the rectangle. Geometric intersections are used to determine the corner positions at high accuracy, making it possible to calculate the pose of the marker with respect to the camera up to a scale factor. If the size of the marker is linked to its ID, the 3D camera pose can be accurately tracked in real-time. Negative points are that the markers must be put precisely in place in the scene beforehand, and that for markers far away the four corners are too close together to make an accurate pose calculation possible. In this case, setting up an exact grid of markers may overcome this, but setting up such an exact grid is very cumbersome and, moreover, contradicts the requirement to work in a context-free manner.

Due to new algorithms and faster computers, real-time natural feature tracking becomes feasible for pose estimation [6, 12, 18]. Simultaneous localisation and mapping (SLAM) and its adaptations such as EKF-SLAM [6] and FastSLAM [7] are the most popular incremental mapping methods. The current camera pose and the position of every landmark (clearly visible natural feature) are updated with every incoming image. However, these methods suffer from a high computational load due to the data association, and no more than a few dozen points can be tracked and mapped in real-time. Monocular setups suffer from the lack of metric distance and their maps are created up to scale. Se and Lowe [18] used a stereo setup and SIFT [13] features as natural features to track the pose of a mobile robot. They assume a 2D planar motion of their platform assuming it operates on planar surfaces such as floors. However, their assumption is not valid for platforms such as robot arms, and their algorithm is computationally expensive. Klein and Murray [11] proposed a parallel tracking and mapping (PTAM) system in which they split the pose estimation and mapping processes by using a key-frame structure. They also utilised a bundle adjuster to correct the positions of landmarks. *Bundle adjustment* is the refinement of a 3D reconstruction to produce jointly optimal 3D structure and viewing parameter estimates such as the combined 3D feature coordinates, camera poses, and calibrations. Their system shows impressive performance in small spaces such as on office desks, but their system suffers from the usual drawbacks of monocular setups.

Dense 3D map reconstruction is a process that can be performed in parallel to pose estimation. The aforementioned pose estimation methods calculate the 3D location and orientation of a camera at each frame and they build up a sparse 3D map of landmarks that can be used for tracking. Consequently, another software module is necessary that is able to create dense 3D maps and merge them with each other. Many researchers have investigated this field [8, 9, 14, 15]. Multi-view approaches [8, 15] estimate maps by using multiple images of the same scene. These methods are either computationally very expensive or confined to small workspace areas. Also their primary goal is to create very detailed 3D maps which is not always necessary for our applications. In [8, 14], 3D maps estimated using a stereo setup (with known 3D pose) are registered by using various assumptions, such as the planarity of the scene or the visibility of points depending on the context. However, these assumptions are not always valid in our application.

13.3 System Design

For the acquisition and reconstruction of 3D environments, apart from multiple camera setups (such as stereo-vision), various other 3D sensors can be used, such as based on laser scanning or time-of-flight (TOF) cameras. However, these systems usually introduce extra weight, power consumption, and cost. Due to the laptop and smart phone developments, CMOS cameras became low-cost, high-quality, commodity products. Therefore, simple CMOS colour cameras are used as the vision sensors and no additional hardware was used to satisfy the simplicity requirement. Two Microsoft LifeCam HD-5000 cameras are used to build a stereo pair with low price, good low-light performance, and reasonably good optics. Moreover, the weight of the circuit board together with its lenses is small, its size is small, and it can be easily mounted on a robot or a mobile platform. The cameras can output colour images with 1280×720 pixel resolution at 30 frames per second (fps). The cameras are mounted approximately 6.5 cm from each other (as shown in Fig. 13.1) and are calibrated off-line to find the rotation and translation between the camera sensors (extrinsic parameters) as well as the camera matrices (intrinsic parameters). Synchronisation of the cameras for the acquisition of the stereo images is achieved in software by applying a multi-threaded design, in which each camera has its own thread for image acquisition.

To fulfil the requirements of context-free pose estimation and mapping system (see Sect. 13.2), a system as depicted in Fig. 13.2 has been developed. This system consists of four main components: a tracker (TR), a sparse map maker (SMM), a dense map maker (DMM), and a map integrator (MI). The benefit of such a modular system is that, in order to cope with new tasks or hardware setups, one can modify and update every part of the system largely independent of the other parts. The 3D pose of the camera is calculated by the TR module in real-time by tracking the landmarks in a 3D sparse map which is created by the SMM module. The DMM module uses the stereo system to create dense local 3D maps from captured video frames. Finally, the MI module uses the pose information and local 3D maps to calculate a global dense 3D map of the scene in near real-time.

Fig. 13.1 Webcam-based stereo setup

13.3.1 Tracker and Sparse Map Maker

Robust real-time pose estimation is one of the most crucial parts of a localisation system, since the 3D pose of the camera in the physical world is necessary to manipulate objects correctly on required positions. Marker-based localisation systems are limited to pose estimation when a marker is in view. Moreover, marker placement is often not possible, allowed, or feasible. Natural features, such as edges and corners, can eliminate the limitations of marker-based systems. These natural features can be used for simultaneous localisation and mapping. Since natural features are already available in the scene, the user can start an application without any prior scene modification. Also the availability of many natural features in a scene makes the system robust, since the tracking can continue as long as some features are visible while others may be occluded.

Our framework uses an improved version of PTAM (Parallel Tracking and Mapping) [11] for pose estimation based on natural features. The PTAM code is improved by replacing the single-camera setup by a stereo-camera setup to enable 3D natural feature matching and multi-camera tracking. A stereo-camera setup enables the system to measure the metric distances between the camera and the detected natural features. Such a setup improves the tracking performance by combining two views and makes "metric map generation" possible.

Various feature types such as SURF [3] and SIFT [13] to extract map points have been implemented. However, their computational load increases the time complexity of the tracker and decreases the robustness against fast motions. Therefore, ultimately the FAST feature detector [16] has been used to detect salient points in the scene. Typical features extracted from the scenes are shown in Fig. 13.3.

When the system starts up, the most salient features are matched between the left and right views, 3D positions are calculated and added to the map by the SMM module. Subsequently, features in each new frame are extracted and 3D features in the map are projected back to the new image by using the previous camera pose. Then, the new features and the back-projected ones are matched and the new position of

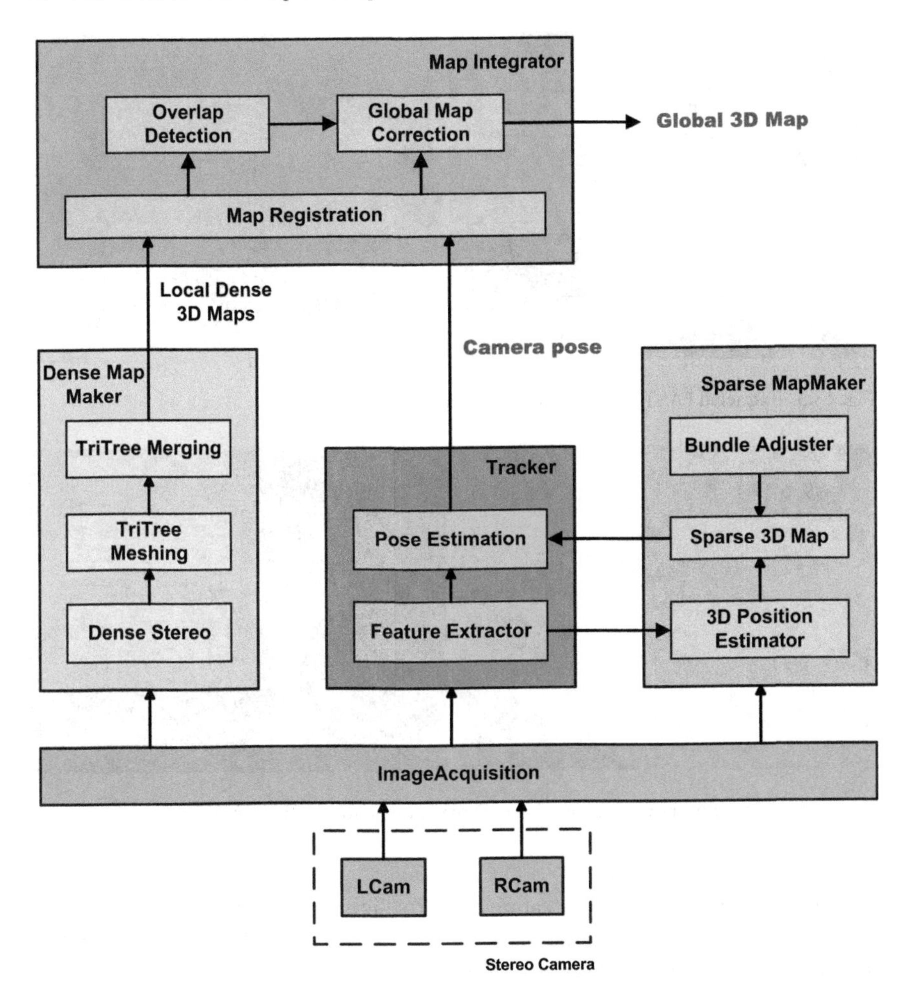

Fig. 13.2 System design of self-localisation and map building

the camera is calculated. This process is repeated for each camera and the calculated 3D poses of the cameras are combined. Each camera contributes to the final pose weighted by the quality of its estimate, calculated by back-projection.

In an ideal application, the robot agent (AGV or manipulator) moves around the scene to interact with the objects in the scene. Therefore, when the user explores new parts of the scene, new landmarks must be added to the set of already tracked features in the coarse map in order to keep the pose estimation correct and continuous. The SMM module extracts new 3D landmarks, in a similar way as in the start-up phase, when the camera moves 10–15 cm away from the last pose that was extracted. Extracted features are compared with the existing ones. Identical features as well as very near features (within 5×5 pixel distance) are discarded, while the rest of the

Fig. 13.3 Extracted FAST features, shown in purple

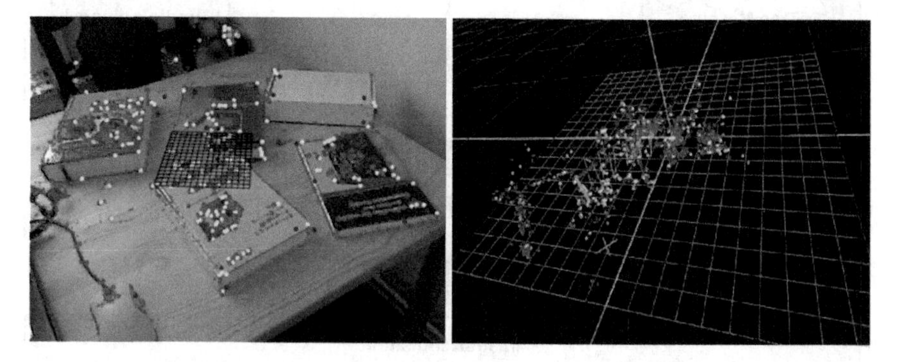

Fig. 13.4 *Left*: 2D back-projections of the 3D map points. *Right*: 3D map points

features are added to the sparse feature map. To eliminate the error in pose estimation and to register new points to the map, bundle adjustment [19] is performed and the pose information and the 3D coordinates of the points are corrected. The total set of tracked 3D features constitutes a sparse map of the scene and can be utilised as an initial map for further processing.

Utilising two cameras for natural feature tracking does not only provide us with metric information due to its stereo set-up, but also increases the robustness of the system. The final pose of the camera pair is estimated by combining the individual estimates of both cameras, therefore the final pose is found as long as one of the cameras is able to track the features.

The 3D map points maintained by the SMM module and their back-projections are shown in Fig. 13.4. The different colours represent the different scales on which the features are extracted.

Our system was tested on an Intel Pentium 4 computer with a 2.2 GHz processor and a Linux operating system. It extracts approximately 800–1,000 features from a normal image with sufficient texture in it in approximately 3 ms including the colour conversion, and it can generate a sparse map of up to 30,000 features. This number of features is enough to represent a room or an office space in indoor applications.

The system can track 2,000 visible features at 35–40 fps. This is more than necessary as approximately 600 features are enough to represent a view. The number of features and the map size can be increased by using a faster computer. The accuracy of the pose estimation is within 1 cm resolution up to 4 m; a comparison of the single camera version with EKF-SLAM on a synthetic sequence is given by Klein and Murray [11].

13.3.2 Dense Map Maker and Map Integrator

A dense 3D map of a scene contains important information for further processes such as shape extraction, obstacle detection, object/human segmentation, path planning, etc. Global dense map generation can be separated into two main modules: local map generation and map registration. The first module is responsible for creating a local dense depth map from incoming stereo images and correcting them, while the second module integrates new local maps within the global map.

The DMM module creates *disparity images*—the images that contain the position differences of similar points in right and left image—and improves them by smoothing or replacing uncertain measurements by using their spatial neighbours. The TR module tracks the 3D location of the stereo-camera setup continuously and stores the incoming stereo images with their poses when the camera displacement with respect to the previously stored image pair is greater than some threshold. Therefore, the frames are only processed when there is some motion. Initial disparity maps are calculated from these image pairs by utilising the Semi Global Matching (SGM) algorithm [10] and a continuous stream of disparity maps are generated when the camera moves around the scene. However, the errors in the disparity maps cause non-overlapping regions and non-closed surfaces. Also because of the narrow baseline and the stereo matching problems, wavy patterns appear when low-textured surfaces are matched.

To overcome these drawbacks and correct the 3D maps we used an algorithm of Sarkis et al. [17], which was originally developed for image compression. Initially the disparity image is divided into two triangles. For each triangle, the plane passing through its three vertices is calculated. Then the pixels inside the triangle are projected onto the plane and the distance between the original points and their projections are calculated. The sum of all distances is used as cost function and the triangle is separated into two if the cost is greater than some threshold. This division continues until there is no triangle with high disparity variation found. The final structure is called a *tritree* [17] and triangles inside the tree represent homogeneous regions. The recursive subdivision is called *meshing*. Following this meshing step, an integration step is performed in which triangles that are similar to each other are combined. Similarity is defined as the spatial proximity of triangles and the angle between their surface normals. The proposed method segments the disparity maps into homogeneous regions while preserving the edges. We also obtain a planar approximation of the scene which can be used to extract shapes. Finally, the disparity values are replaced by their projections onto the triangles to eliminate noise and uncertain disparities.

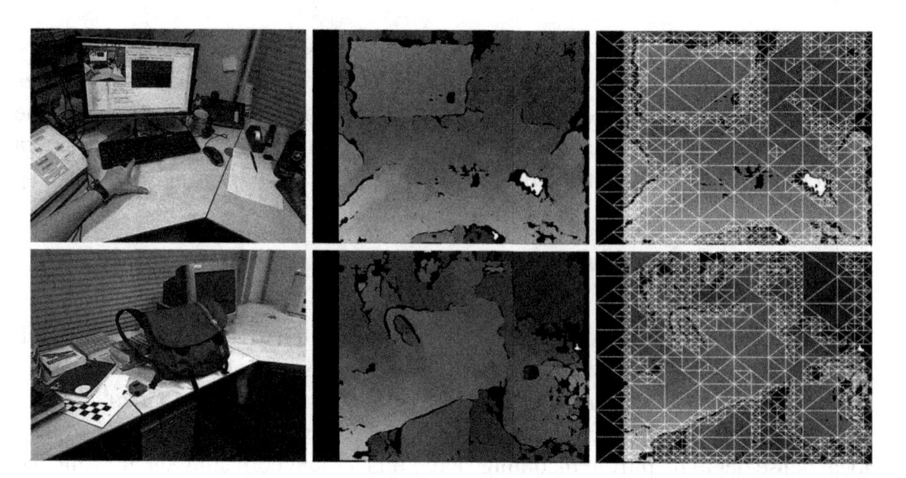

Fig. 13.5 *Left*: input (left camera) images. *Middle*: calculated disparity maps. *Right*: tritree structures

This method has near-real time performance (approximately 5 frames per second for 2,000 triangles). Note that the tritree structure is only created when there is a significant motion of the cameras and this relaxes the real-time requirements. Typical tritree structures are shown in Fig. 13.5.

Corrected local disparity maps are combined to create a global map. The TR module provides the pose of every stereo pair in 3D space. This information is used to combine local maps by the MI module. The coordinate frame derived from the first pair of stereo-images is selected as the main coordinate frame and all points coming from further stereo image pairs are projected to that main coordinate frame. However, because of the pose estimation errors, local maps do not fit perfectly. Therefore, the triangles in the overlapping regions are compared and the pose estimates are corrected. The global maps created in this way are presented in Fig. 13.6.

13.4 Conclusion and Future Work

In this chapter, a novel framework for pose tracking and map building for collision-free robot and Automated Guided Vehicle (AGV) motion in context-free environments is presented. Robotic manipulation for order picking is one of the challenges for future warehouses. In every phase of the process (object detection and recognition, object grasping, object transport, and object deposition), avoiding collisions is crucial for successful operation. This is true for different warehouse designs, i.e. man-to-goods and goods-to-man types of systems, and robot arms and AGVs: they all need to know their 3D pose within the 3D map of their surroundings to perform their task, using path planning and/or visual servoing. On-line map generation provides the possibility to adapt to dynamic environments.

Fig. 13.6 Created global dense 3D maps from seven stereo images

A system has been designed and implemented that can track in real-time the pose of a stereo pair of cameras that can be mounted on a robot manipulator or AGV. This 3D pose of the system is tracked by using natural features that are observed in a scene. The 3D pose of the features with respect to the camera pair (and vice versa) are calculated by using stereo vision and new landmarks are added to a coarse 3D feature map as the camera moves. A sub-set of the stereo images are also used to create a global dense map of the environment after correction for noise and views with sparse features. The output of our system is both a sparse map that is used to provide the pose of the camera in real-time with an accuracy of about 1 cm when viewing natural features within 4 m of the camera and a dense map—a point cloud—that can be used to fit principle planes like ground planes, walls, ceiling, table surfaces etc. onto it, in order to generate a geometric description of the space. (Ground) plane fitting has been achieved but is not yet integrated into the system. More elaborate measurements on the camera pose accuracy and its dependence on system parameters and environment are yet to be performed.

As future work, first of all the dense map can be used to recover the fixed geometry of the surrounding space. Note that CAD-drawings of this space and of man-made objects may help the geometric reconstruction. Then, moving objects in the scene, e.g. humans, can be found using the disparity images or the differences in successive dense maps and—described with a 3D convex hull—be used with the fixed geometric structure of the space to perform path planning in. Second, the method can be combined with the systems presented in Chaps. 10 and 11. Furthermore, our real-time pose estimation and map making can be combined with various other computer vision algorithms to extend the application field [1, 2, 4].

References

1. Akman O, Bayramoglu N, Alatan AA, Jonker P (2010) Utilization of spatial information for point cloud segmentation. In: 3DTV-conference: the true vision—capture, transmission and display of 3D video (3DTV-CON), pp 1–4
2. Akman O, Jonker P (2010) Computing saliency map from spatial information in point cloud data. In: Advanced concepts for intelligent vision systems. Lecture notes in computer science, vol 6474, Springer, Berlin, pp 290–299
3. Bay H, Ess A, Tuytelaars T, Van Gool L (2008) Speeded-up robust features (SURF). Comput Vis Image Underst 110:346–359
4. Bayramoglu N, Akman O, Alatan AA, Jonker P (2009) Integration of 2D images and range data for object segmentation and recognition. In: Proceedings of the twelfth international conference on climbing and walking robots and the support technologies for mobile machines, pp 927–933
5. Caarls J, Jonker P, Kolstee Y, Rotteveel J, van Eck W (2009) Augmented reality for art, design and cultural heritage—system design and evaluation. EURASIP J Image Video Process. Article ID 716,160
6. Davison AJ, Reid ID, Molton ND, Stasse O (2007) MonoSLAM: Real-time single camera SLAM. IEEE Trans Pattern Anal Mach Intel 29:1052–1067
7. Eade E, Drummond T (2006) Scalable monocular SLAM. In: 2006 IEEE computer society conference on computer vision and pattern recognition, pp 469–476
8. Gallup D, Frahm JM, Pollefeys M (2010) Piecewise planar and non-planar stereo for urban scene reconstruction. In: 2010 IEEE conference on computer vision and pattern recognition (CVPR), pp 1418–1425
9. Habbecke M, Kobbelt L (2007) A surface-growing approach to multi-view stereo reconstruction. In: IEEE conference on computer vision and pattern recognition, CVPR '07, pp 1–8
10. Hirschmuller H (2008) Stereo processing by semiglobal matching and mutual information. IEEE Trans Pattern Anal Mach Intel 30:328–341
11. Klein G, Murray D (2007) Parallel tracking and mapping for small AR workspaces. In: 6th IEEE and ACM international symposium on mixed and augmented reality, ISMAR 2007, pp 1–10
12. Lee YJ, Song JB (2009) Visual SLAM in indoor environments using autonomous detection and registration of objects. In: Multisensor fusion and integration for intelligent systems. Lecture notes in electrical engineering, vol 35. Springer, Dordrecht, pp 301–314
13. Lowe DG (2004) Distinctive image features from scale-invariant keypoints. Int J Comput Vis 60:91–110
14. Merrell P, Akbarzadeh A, Wang L, Mordohai P, Frahm JM, Yang R, Nister D, Pollefeys M (2007) Real-time visibility-based fusion of depth maps. In: IEEE 11th international conference on computer vision, ICCV 2007, pp 1–8

15. Newcombe RA, Davison AJ (2010) Live dense reconstruction with a single moving camera. In: 2010 IEEE conference on computer vision and pattern recognition (CVPR), pp 1498–1505
16. Rosten E, Drummond T (2006) Machine learning for high-speed corner detection. In: Computer vision—ECCV 2006. Lecture notes in computer science, vol 3951. Springer, Berlin, pp 430–443
17. Sarkis M, Zia W, Diepold K (2010) Fast depth map compression and meshing with compressed tritree. In: Computer vision—ACCV 2009. Lecture notes in computer science, vol 5995. Springer, Berlin, pp 44–55
18. Se S, Lowe D, Little J (2002) Mobile robot localization and mapping with uncertainty using scale-invariant visual landmarks. Int J Robot Res 21:735–758
19. Triggs B, McLauchlan P, Hartley R, Fitzgibbon A (2000) Bundle adjustment—a modern synthesis. In: Vision algorithms: theory and practice. Lecture notes in computer science, vol 1883. Springer, Berlin, pp 153–177

Chapter 14
Flexible Transportation in Warehouses

Sisdarmanto Adinandra, Jurjen Caarls, Dragan Kostić, Jacques Verriet, and Henk Nijmeijer

Abstract In recent years, autonomous mobile robots (AMR) have emerged as a means of transportation system in warehouses. The complexity of the transport tasks requires efficient high-level control, i.e. planning and scheduling of the tasks as well as low-level motion control of the robots. Hence, an efficient coordination between robots is needed to achieve flexibility, robustness and scalability of the transportation system. In this chapter, we present a methodology to achieve coordination in different control layers, namely high-level and low-level coordination. We investigate how the coordination strategies perform in an automated warehouse. We use simulation results to analyse the system performance. We take into account typical performance indicators for a warehouse, such as time to accomplish the transportation tasks and total cost of the system. In addition to the simulation results, we conduct experiments in a small-scale representation of the warehouse.

S. Adinandra (✉) · J. Caarls · D. Kostić · H. Nijmeijer
Department of Mechanical Engineering, Eindhoven University of Technology,
P.O. Box 513, 5600 MB Eindhoven, The Netherlands
e-mail: s.adinandra@tue.nl

J. Caarls
e-mail: j.caarls@tue.nl

D. Kostić
e-mail: d.kostic@tue.nl

H. Nijmeijer
e-mail: h.nijmeijer@tue.nl

J. Verriet
Embedded System Institute, P.O. Box 513, 5600 MB Eindhoven, The Netherlands
e-mail: jacques.verriet@esi.nl

R. Hamberg and J. Verriet (eds.), *Automation in Warehouse Development*,
DOI: 10.1007/978-0-85729-968-0_14,

14.1 Introduction

The transportation in warehouses is typically organised using conveyor systems. These systems provide a good transport capacity and high availability, but are sensitive to conveyor failure. In the event of a conveyor breakdown, the transport system of the warehouse will likely to come to a complete standstill. This is due to the fact that their relative high cost in the warehousing industry restricts having too much redundancy in the transportation system. Another weakness is their fixed maximum capacity. If the business process of the warehouse changes, a larger capacity may be needed. Once the conveyor capacity is exceeded, large changes in the warehouse layout are needed to accommodate the changed business process. Besides having a high performance, an ideal warehouse transport system should be robust to system failures and flexible in to system changes. An autonomous mobile robot (AMR) transport concept has these desired characteristics [14].

In the AMR concept, a large collection of autonomous mobile robots, called *shuttles*, is responsible for the transportation of goods in a warehouse. The robustness and flexibility of an AMR system are apparent. As the system has a large transport redundancy, the breakdown of a single robot may lower the system's performance, but will not lead to a complete system standstill. Flexibility can easily be achieved by varying the number of robots in the system: by adding or removing robots, an AMR transport system is capable of handling variation in transport demand, either due to seasons or business changes.

Since robots offer more flexibility than traditional conveyors, the control of an AMR transport requires a more sophisticated approach so that a performance similar to the conveyor systems, or an even better performance, can be achieved. This involves control of an individual robot, like the assignment of transport tasks to robots, but also inter-robot coordination, like avoidance of collisions and gridlocks. Another challenge involves the robots' awareness of their environment. In order to function optimally, robots should be able to observe and reason about their environment and change their behaviour accordingly.

In transportation system, planning, scheduling, and control of tasks of the conveyors or robots are done by a high-level control system in a centralised or decentralised way, see e.g. Weyns et al. [13] and Gu et al. [5]. The high-level control system allocates the tasks to the robots based on customer orders and resource availability. The centralised strategy typically produces the optimal throughput in the absence of uncertainties. However, it exhibits some weaknesses in the case of unexpected events that affect the system, such as time-consuming replanning, some robots block other robots' paths, or start to deviate from their assigned path. The decentralised strategy results in a suboptimal throughput, but is more robust against disturbances.

Recent developments in formation and coordination control of mobile robots by low-level motion controllers show promising results. Various techniques, e.g. leader-follower and virtual structure, can be used to realise the transportation system. In the low-level approach, the motion controller achieves tracking of individual robot trajectories and maintains the desired spatial formation between the robots. See for instance, Arai et al. [2], Chen and Wang [3], and Liu et al. [11] for reviews and recent developments about the low-level motion coordination.

In this chapter, we provide information on how to control a large collection of autonomous mobile robots such that all performance requirements are met at an acceptable cost. We show two different coordination algorithms, namely high-level and low-level coordination. We analyse the performance of the AMR system in an automated warehouse and compare it to the warehouse's conveyor system. We investigate the cost and performance of the proposed system.

This chapter is organised as follows. In Sect. 14.2, we describe our control architecture, the kinematic model of the mobile robots and its trajectory tracking controller, as well as the performance indicators used in this chapter. Section 14.3 explains in detail the control strategies to coordinate the robots. Section 14.4 highlights the automated warehouse. Section 14.5 reports the simulation and experimental results and highlights the main findings of this chapter. Conclusions and future work are presented in Sect. 14.6.

14.2 Preliminaries

This section explains the control architecture, the mobile robots model and the trajectory tracking used in this chapter. The performance indicators used for evaluation is introduced in this section as well.

14.2.1 Control Architecture

Control design and task planning in a warehouse are complex and difficult tasks. These involve design of warehouse layout, choice of storage type, storage replenishment, order arrangement, task dispatching, capacity of conveyors, number of workstations, etc., see e.g. Gu et al. [5]. Narrowing the problem to transportation only does not reduce the complexity. In conveyor systems, it includes complex task planning based on the customer order and conveyor capacity. On the other hand, in an AMR system the control and planning challenges include task scheduling, dispatching rules, and motion control of the robots. Typically these problems are solved either as one, or separate design problems.

In an AMR-like system, like the one presented by Lacomme et al. [10], one tries to combine the problem of task scheduling and robots dispatching in one optimisation problem. In this chapter, we choose to use the separation approach. We decompose the control and task planning into different control layers as shown in Fig. 14.1. The control layers give us the convenience of having the control design isolated from the rest of the system and the possibility to test different control algorithms. These layers also allow shifting responsibilities between the control tasks.

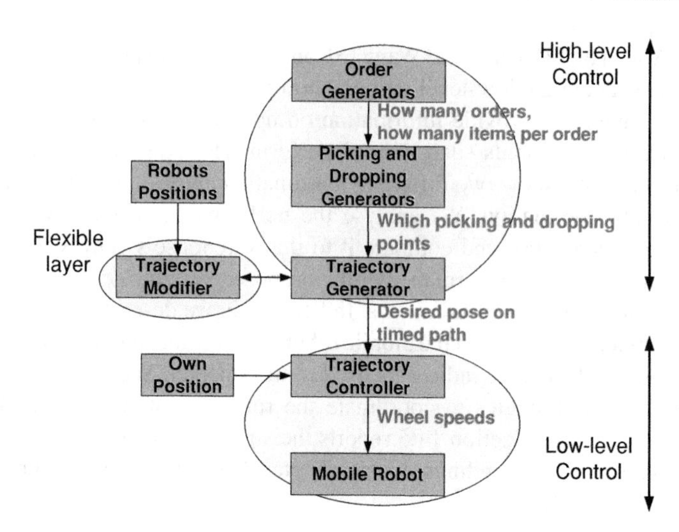

Fig. 14.1 Proposed control architecture

- High-level control
  - *Order generators.* The main task of this layer is to generate orders. Additionally, it decides which items and how many of them are needed in each order.
  - *Picking and dropping generators.* Based on the generated orders, this layer determines to which pick-up/drop-off points robots have to go.
  - *Trajectory generators.* This layer is responsible for generating reference trajectories of the robots. These trajectories may or may not allow occurrence of collisions.
- Low-level control
  - *Trajectory controller.* This layer is responsible for accurate tracking of the reference trajectories.

Figure 14.1 also shows a *flexible layer* which can be made part of the high-level or low-level control. This layer is to show how we can shift responsibilities between the layers. In this chapter, we show an example of shifting collision avoidance responsibilities between the high-level and low-level control.

14.2.2 Unicycle Mobile Robots

There are various types of autonomous vehicles. Some of them belong to the class of vehicles with non-holonomic constraints. In this particular case, it means that the total number of degrees of freedom is larger than the number of controllable degrees of freedom. In practice, this constraint implies that no sideways movement is allowed.

Throughout this chapter, we consider a group of m autonomous mobile robots that are described by the non-holonomic kinematic model of a unicycle mobile robot (see e.g. Kanayakama et al. [7] and Jiang and Nijmeijer [6], and references therein):

$$\begin{bmatrix} \dot{x}_i \\ \dot{y}_i \\ \dot{\theta}_i \end{bmatrix} = \begin{bmatrix} v_i \cos \theta_i \\ v_i \sin \theta_i \\ \omega_i \end{bmatrix}. \tag{14.1}$$

Here, v_i and ω_i are the forward and angular velocities, respectively. x_i and y_i are the Cartesian coordinates of the robot centre in the world coordinate frame. θ_i is the heading angle relative to the x-axis of the world frame, and $i \in \{1, 2, 3, \ldots, m\}$.

14.2.3 Trajectory Tracking

The velocities v_i and ω_i in Eq. 14.1 are two control inputs that we can use to control the robot movement. In most existing AMR-like systems, the robots follow fixed paths, i.e. only controlling v_i. In this solution, the high-level control assigns fixed trajectories that results in optimal throughput, see Lacomme et al. [10] as an example. This approach is simple, but cannot easily accommodate changes in transportation demands. Thus, we propose to use a trajectory tracking controller. The idea is as follows. The high-level control provides a reference trajectory to each robot:

$$\mathbf{p}_{ri} = \left[x_{ri} \; y_{ri} \; \theta_{ri} \right]^T. \tag{14.2}$$

The geometry of the reference trajectory is free to choose as long as is fulfils the non-holonomic constraints, i.e. $-\dot{x}_{ri} \sin \theta_{ri} + \dot{y}_{ri} \cos \theta_{ri} = 0$. The low-level control is responsible for accurate tracking of these reference trajectories. In this chapter, we use the following controller:

$$v_i = v_{ri} \cos \theta_{ei} + k_{xi} x_{ei}, \tag{14.3a}$$

$$w_i = w_{ri} + k_{yi} v_{ri} y_{ei} \frac{\sin \theta_{ei}}{\theta_{ei}} + k_{\theta i} \theta_{ei}, \quad \text{with} \quad \frac{\sin \theta_{ei}}{\theta_{ei}} = 1 \text{ if } \theta_i = 0. \tag{14.3b}$$

Here, v_{ri} and ω_{ri} are the reference forward and angular velocities, respectively. x_{ei}, y_{ei}, and θ_{ei} are the tracking errors represented in robot local coordinate frame [7], and $k_{xi}, k_{yi}, k_{\theta i} \in \mathbb{R}^+$ are control gains. The controller in Eqs. 14.3a, b is a modified version of the controller proposed by Jiang and Nijmeijer [6].

14.2.4 Performance Indicators

To get insight into the behaviour of the proposed transportation system and to evaluate its performance, we use the following indicators:

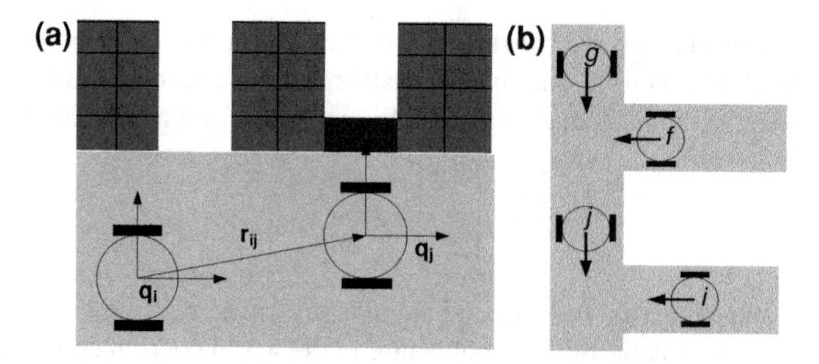

Fig. 14.2 An example where two robots almost collide: **a** situation in which robot j stops. Robot i has to take an action to avoid collision; **b** situation with robots at two junctions

- Completion time, t_{complete}, which is the time needed to accomplish all transportation tasks. A smaller value of completion time means that more tasks can be handled in a fixed period of time. This also means a higher throughput of the warehouse. The completion time is computed as follow:

$$t_{\text{complete}} = t_{\text{last,task}} - t_{\text{first,task}}, \tag{14.4}$$

  where $t_{\text{first,task}}$ and $t_{\text{last,task}}$ are the times for starting the first and completing the last task, respectively.
- Total cost, $\text{cost}_{\text{total}}$. This is to measure the cost of the transportation system.

14.3 Coordination Strategies

In this section, we discuss how the control architecture allows different control strategies to be implemented in the flexible layer shown in Fig. 14.1. Here we demonstrate how collision avoidance is solved in two different control approaches, i.e. high-level and low-level coordination. A typical example where collision can occur is shown in Fig. 14.2.

The collision threat seen in Fig. 14.2 can be solved either by making robot i slow down/stop by modifying the reference velocities or by creating a new non-colliding path. The next two sections explain how slowing down can be done in two different control layers using two different algorithms.

14.3.1 High-Level Coordination

The proposed high-level collision avoidance algorithm works on the level of the trajectory generator. In this case, the desire timed path for each robot is already fixed using nominal velocity. The actual velocity has to be adapted to avoid collisions and to get back on track when robot is behind schedule.

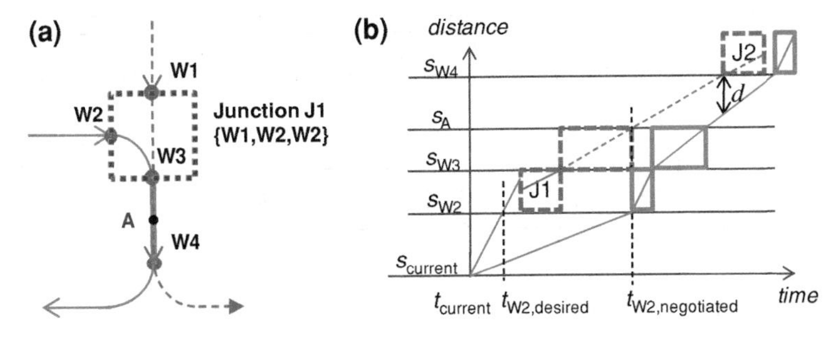

Fig. 14.3 An illustration of the high-level coordination. **a** An example with four waypoints and the paths of two robots; **b** desired/negotiated trajectory of a vehicle (*solid*) and occupation intervals of other higher priority vehicle (*dashed*)

On this high-level coordination, the robots communicate intended arrival times at possible collision points, i.e. junctions, as well as their entry and exit direction for each junction. Using that information, the high-level coordination algorithm tries to simultaneously maximise throughput at each junction and minimise the waiting time for each robot.

The high-level coordination algorithm maximises the throughput by making the robots move quickly over the junction without slowing down. As a consequence, they do not block the passage for robots wanting to cross from another direction. This means that robots only should enter a junction when they are able to leave it. The waiting time is minimised by giving a robot that is more behind schedule a higher priority to be on the junction. Therefore, this delay has to be communicated as a priority value. This method prevents starvation. During normal operation, robots arriving from multiple directions are allowed alternately on the junction.

To make negotiations efficient, the robots only communicate about the previous and next junction they pass(ed), and only do so when needed. When a robot has crossed a junction, it will communicate its intended arrival time at the next junction, and if other robots think they have priority, they will react. This starts a message stream until they all agree. The robots adapt their velocities and some may have to update their departure time for the area after the passed junction (passing time of point A in Fig. 14.3). These messages are broadcast to all robots going towards the involved junction, so one message reaches all interested robots. A decentralised publish/subscribe mechanism is used to keep track of interested robots.

In addition, each robot locally caches the schedule for the next junction, so they can anticipate the reaction of the other robots, preventing the selection of many wrong arrival times. This keeps the number of messages at a minimum. Only to cope with errors in communication, one could set a maximum time between broadcasts of arrival times, e.g. in our simulations we used 0.5 s. The high-level coordination algorithm is summarised in Fig. 14.4.

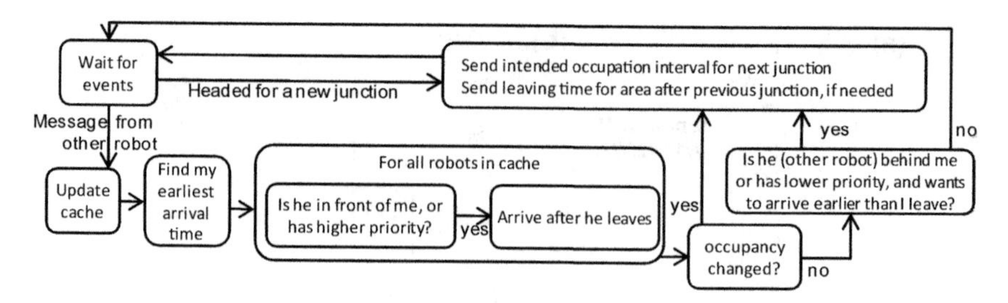

Fig. 14.4 Simplified flow diagram of the high-level coordination algorithm loop in each robot

14.3.2 Low-level Coordination

The second collision avoidance algorithm is implemented at the low-level control, i.e. the flexible layer is shifted to low-level control. The low-level control is mainly responsible to accurately track the reference trajectories given in Eq. 14.2 using the controller in Eqs. 14.3a, b. In high-level coordination, the states (trajectories and arrival times at junctions) of all robots are shared between robots. In low-level coordination, the collision avoidance is solved by using the positions of other robots in the neighbourhood that can be detected using a robot's distance sensor. We implement local coordination that is a modified version of the algorithm of Kostić et al. [9].

Let us recall the situation shown in Fig. 14.2. If $\mathbf{q}_i = \left[x_i \; y_i \; \theta_i \right]^T$ and $\mathbf{q}_j = \left[x_j \; y_j \; \theta_j \right]^T$ are the position and orientation of robots i and j in Cartesian space, we define the vector

$$\mathbf{r}_{ij} = \left[x_j - x_i \; y_j - y_i \right]^T, \tag{14.5}$$

with its magnitude, representing the distance between the centres of robots i and j:

$$|\mathbf{r}_{ij}| = \sqrt{(x_j - x_i)^2 + (y_j - y_i)^2}. \tag{14.6}$$

Define the projection of the direction of robot i: $\mathbf{dir}_i = \left[\cos\theta_i \; \sin\theta_i \right]^T$. The slow-down coefficient of robot i with respect to robot j is expressed as:

$$\sigma_{ij} = \begin{cases} 1, & \text{if } \mathbf{dir}_i \cdot \mathbf{r}_{ij} \le 0 \\ \delta_{\gamma ij}\left(|\mathbf{r}_{ij}|\right) & \text{if } \mathbf{dir}_i \cdot \mathbf{r}_{ij} > 0 \end{cases}, \tag{14.7}$$

where the $\cdot$ sign represents the dot product of two vectors and $\delta_{\gamma ij}\left(|\mathbf{r}_{ij}|\right)$ is a penalty function of Kostić et al. [9]. The reference forward velocity of each robot is penalised as follows

$$v_{ri} = v_{\text{des},i} \prod_{j=1, j\neq i}^{m} \sigma_{ij}, \tag{14.8}$$

where $v_{des,i}$ is the desired forward velocity of each robot and m is the number of robots.

The coefficient computed in Eq. 14.7 indicates whether robot j is behind or in front of robot i relative to the direction of robot's i movement. According to Eq. 14.8, if robot i is behind robot j and the other robots, then robot i will slow down to avoid collisions.

Furthermore, priority rules need to be applied if the situation shown in Fig. 14.2b occurs, i.e. more robots wait to enter the junction. Applying only Eq. 14.8 will result in deadlock, i.e. no robots move. This is because robots that enter the junction assume they are behind each other, i.e. $\sigma_{fg} = \sigma_{gf}$ and $\sigma_{ij} = \sigma_{ji}$. Suppose we implement right-hand priority, i.e. at the junction a robot that comes from the right side of other robots has higher priority, $\sigma_{fg}, \sigma_{gf}, \sigma_{ij}, \sigma_{ji}$ is adapted as follows

$$\begin{aligned} \sigma_{ji} &= 1, \ \ \sigma_{ij} = 0, \\ \sigma_{fg} &= 0, \ \sigma_{gf} = 1. \end{aligned} \tag{14.9}$$

The value of $\sigma_{ji} = 1$ and $\sigma_{gf} = 1$ mean that robots j and g have higher priority than robots i and f, respectively. Other simple rules like lower/higher robot id or left-hand priority can be used to replace the right-hand priority rule.

Remark. Slowing down combined with priority rules presented above does not guarantee a deadlock-free evolution if the number of robots exceeds the available space for movement.

14.3.3 Collision Avoidance Using Artificial Potential Fields

The two coordination algorithms presented in the previous two sections are used in normal situation, i.e. there is no fault in the system and all robots all operational. To add robustness against faults, i.e. some robots are subject to failure or unexpected obstacles block the paths, we add collision avoidance using the artificial potential field (APF) algorithm presented by Kostić et al. [8] both to high-level and low-level coordination. In the APF algorithm, a robot generates a repulsive force based on other robots' positions. In this way, a robot can alter its path and avoid collisions with broken-down robots.

14.4 A Case Study in an Automated Warehouse

In this section, we briefly discuss an automated warehouse for the case study as well as assumption on the specification of the robot.

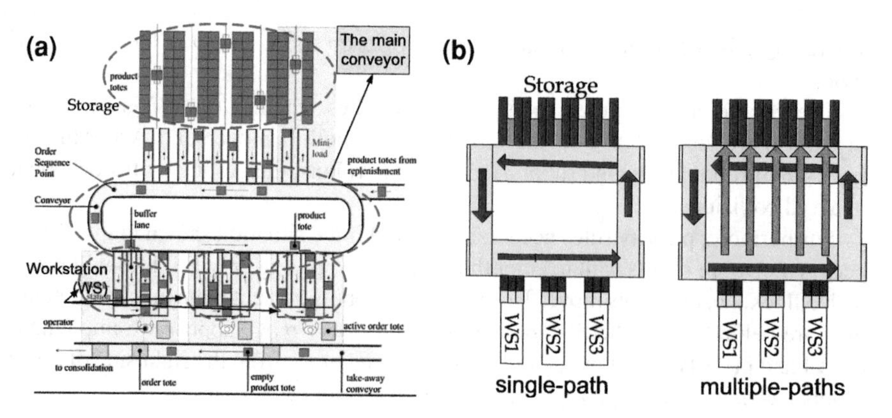

Fig. 14.5 **a** An automated warehouse [1]; **b** two possible paths for the robots

14.4.1 An Automated Warehouse

In simulations, we investigate the performance of the transportation system of the automated warehouse shown in Fig. 14.5 [1]. It is an automated item-picking warehouse. The warehouse is also similar to the one modelled in Chap. 5 of this book.

In the warehouse, human operators (pickers) are responsible for order completion. An order consists of several items that determines the order size and may vary significantly. Items of the same type are transported using a product tote from storage to order-picking workstations. At the workstations, these totes create queues on buffer conveyors. Once the picker and the required product totes are available, the totes will be moved from the buffer to the pick location where the picker stands for order completion.

14.4.2 Using AMR to Replace Conveyors

We deploy autonomous mobile robots to replace the main conveyor loop, i.e. the conveyor that transports product from the storage to the workstations as shown in Fig. 14.5a. The task of a robot is to transport product totes from storage to buffers/workstations. In this chapter, we consider two possible geometries, i.e. single-path and the multiple-paths illustrated in Fig. 14.5b. The single-path geometry resembles the geometry of the original conveyor systems, while the multiple-paths geometry allows robots to take a shortcut. The results presented in this section are not limited to these two geometries.

14.4.2.1 Robot Specifications

The following assumptions are used for the specification of the robots:

- The size of the robot is 0.8 m × 0.8 m. This is equal the size of the biggest totes used in the automated warehouse.
- The nominal speed of the robot is 1 m/s.
- For each task, a robot carries a product tote containing one type of sku/item.

14.5 Performance Analysis

In this section, we describe the scenario investigated in the simulations, also partly in the experiment, and discuss the main findings of this chapter.

14.5.1 Scenarios

We consider different scenarios. In each scenario, we simulate different number of robots (n_{robot}), priority rules, and fault statuses. Considering the size of the robots and the automated warehouse, we choose $n_{\text{robot}} \in \{2, 4, 6, 8, 10, 12, 14, 16, 18, 20\}$. For the fault case, we simulate a situation where two robots are subject to failure. We assume a broken-down robot can be fixed within 30 minutes after which it becomes operational again. For the basis of comparison, we use the maximum capacity of the conveyor in the automated warehouse, i.e. the conveyor can transport 1,000 totes per hour.

The following abbreviations are used to identify the scenarios: SP-HLC: single-path, high-level control; MP-HLC: multiple-paths, high-level control; SP-LLC: single-path, low-level control; MP-LLC-LN: multiple-paths, low-level control, low number priority; MP-LLC-LH: multiple-paths, low-level control, left-hand priority; MP-LLC-RH: multiple-paths, low-level control, right-hand priority. The conveyor system is identified by CS.

14.5.2 High-level versus Low-level Coordination

Figure 14.6 shows how the high-level and low-level coordination perform. Each curve shows how t_{complete} evolves with different collision avoidance algorithms and numbers of robots. Take an example in scenario MP-HLC, we can observe that by adding extra robots, the completion time is reduced. However, beyond a certain number, in this case 12, the high-level coordination suffers from deadlock. Since high-level coordination involves negotiation of occupation time at any intersecting path

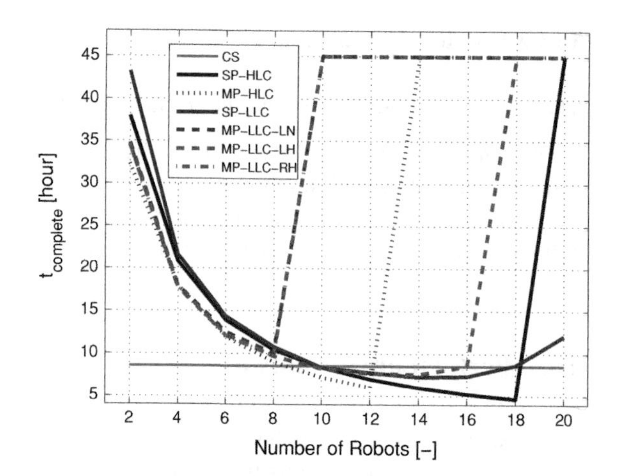

Fig. 14.6 Comparison of different t_{complete} in different scenarios. A value of 45 h means livelock/deadlock (i.e. no robot moves)

segments of all robots, it results in a better throughput than low-level coordination. However, one should remember that low-level coordination does not require (state) information sharing between the robots.

Moreover, Fig. 14.6 indicates how we should choose an appropriate number of robots so that similar throughput as the conveyor system can be achieved, as well as increasing the throughput. Observation of Fig. 14.6 indicates that the appropriate number of robots for this case study is between 8 and 18 depending on the coordination strategies. Above 18, the throughput is lower and the possibility of having a livelock or deadlock increases. Although adding robots can increase the capacity, if the space is kept constant, this also means less space for movement. Thus, the queuing and waiting time of a robot (for the other robots) increases, which result in a higher completion time. Simulation suggests that the combination of 18 robots, single-path, and high-level coordination gives the smallest completion time.

The discussion in the previous paragraph also indicates that using AMR system the scalability of the transport system can easily be increased or decreased by adding or removing robots or by choosing different geometric paths of the robots. There is no additional space needed. Moreover, using the separation principle in the control architecture we can implement different control and planning strategies. This gives us advantage in optimising one layer of the transport system without changing the solution of other layers, e.g. we can focus on optimising the low-level control while keeping all high-level control the same.

Furthermore, Fig. 14.7 shows how the APF algorithm performs in case of failure. As an example, we show how the APF combined with two different low-level coordination scenarios, SP-LLC and MP-LLC-LH, performs. The ability of APF to generate alternative paths for the operational robots makes the overall transport system still operate under failure, albeit with a lower throughput. This means that robustness against failure is achieved. This phenomenon can be observed by the shifting of the original solid-line curves to the dashed-line curves in Fig. 14.7.

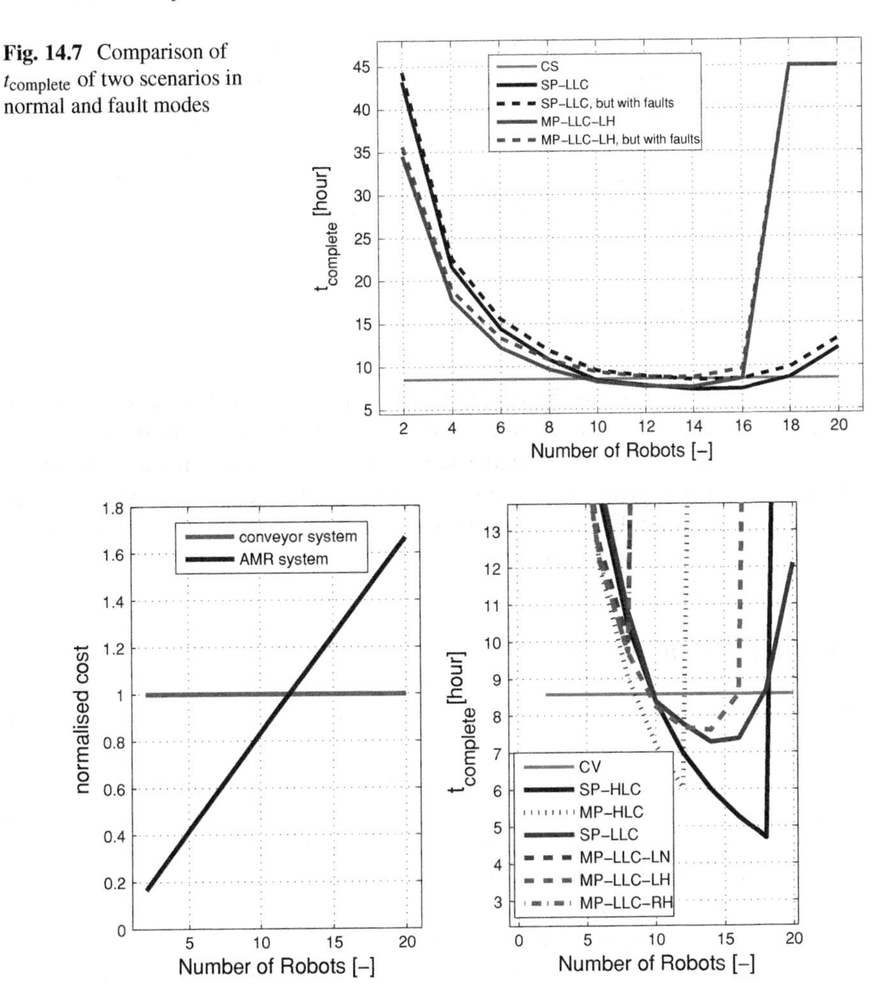

Fig. 14.7 Comparison of t_{complete} of two scenarios in normal and fault modes

Fig. 14.8 *Left:* normalised cost of the transport system. *Right:* comparison of t_{complete} in different scenarios (*zoomed in*)

14.5.3 Cost Analysis

In this section, we discuss the cost comparison between the conveyor and the AMR system. The cost of an AMR system is mainly determined by the number of robots used in the system. We calculate the cost as a normalised value. Any possible cost is normalised to the cost of the conveyor, i.e. $\text{cost}_{\text{total,conveyor}} = 1$. The cost comparison is shown in the left hand side of Fig. 14.8.

From the left-hand side of Fig. 14.8 we observe that for 12 robots the cost of the AMR system is the same as the cost of the conveyor system. As shown in right-hand side of Fig. 14.8, in some scenarios the throughput using 12 robots is better than the conveyor system. Furthermore, with 10 robots we can have similar performance as the conveyor system. It is also easy to observe that for better performance, using more than 12 robots, the AMR cost more with margins up to 50%. Although our cost calculation may not include all components to build the complete transport system, this simple calculation indicates that the AMR has an acceptable cost-versus-performance ratio compared to the conveyor system.

14.5.4 Experimental Validation

In this section, we discuss the experimental validation of the AMR concept. The experiments are done with the experimental setup shown in Fig. 14.9a. The size of the setup is 3.4 m × 2.1 m. A smaller-scale warehouse layout similar to the one in the simulation is implemented. The setup consist of unicycle mobile robots of type e-puck [12] shown in Fig. 14.9b, a two-camera system as seen in Fig. 14.9c, and a PC equipped with a Bluetooth dongle.

14.5.4.1 Experimental Setup

The cameras are used for getting the position and orientation of all robots. All robots are equipped with a home-made marker so the camera system can track the robots. The PC generates robot reference trajectories, processes camera images to get the actual poses of the robots. The PC also computes the tracking control signals as well as the collision avoidance algorithm. The PC sends the control velocities to the robots via a Bluetooth protocol. This way of implementation is chosen because of the limited processing power of the on-board robot processors and due to the limited bandwidth of the Bluetooth communication. Figure 14.9d shows an example of a situation where robots, in this case robots 7–10, have to coordinate their movement to avoid collision and deadlock.

Figure 14.10 shows the history of the control signals and distances between the robots involved in the situation shown in Fig. 14.9d. The data is captured from experiment using MP-LLC-LH.

The experiment results confirm that the strategies work in a practical situation. Videos of the conducted experiments can be seen in [4]. The low-level coordination can handle some uncertainties in typical real-time situation, e.g. noise in position and orientation measurements and small time delays in sending the control signals.

Figure 14.10a shows an example of how the low-level algorithm adapts the control signal to solve the collision threat. In this example, robot 7 stops to pick an item. As a consequence, robot 8 has to slow down to avoid collision, as well as robots 9 and 10. This can be observed in Fig. 14.10a where the control signals of the robots become zero. Once robot 7 starts to move again, robot 8 will start moving. By the left-hand priority, robot 10 gets higher priority than 9, so robot 10 will move forward followed by robot 9. Figure 14.10b shows that the distances (Δ) between robot 7–10 are all

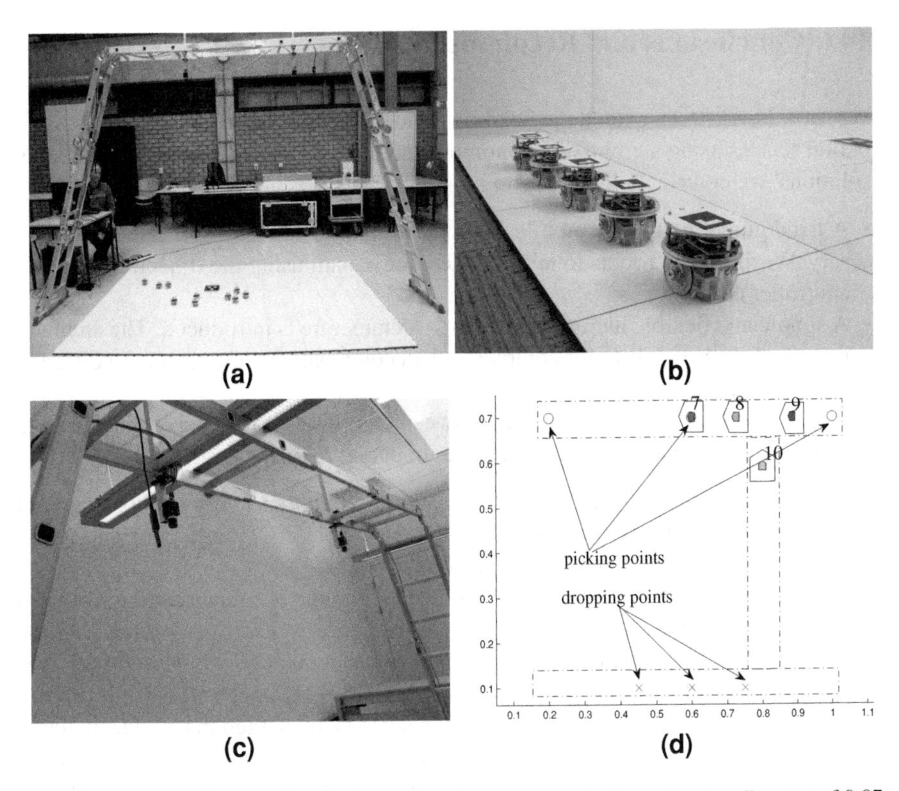

Fig. 14.9 **a** The experimental setup; **b** mobile robot type e-puck. The robot has diameter of 0.07 m; **c** the two-cameras systems; **d** threat of collisions at a junction

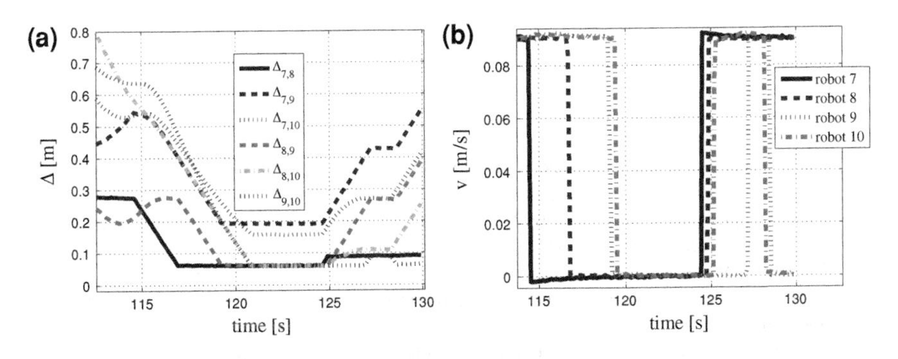

Fig. 14.10 **a** Control signals of robots 7–10; **b** mutual distances between robots 7–10

above 0.06 m, which is the safety distance between the robots. Thus, no collision occurs.

14.6 Conclusions and Recommendations

We show how we can extend the flexibility and robustness in transportation of distribution centres using a group of autonomous mobile robots. Compared to the existing solutions, we contribute the following:

- A fixed guideline for the mobile robot movement is not used and also not necessary. We allow the robots to take any geometric path using the trajectory tracking controller concept.
- A simple and flexible hierarchical control architecture is introduced. The architecture gives the possibility for complex task decomposition, easy testing of different strategies and algorithms.
- We can easily shift responsibilities between control layers (in the architecture) to make different trade-offs between throughput and robustness.

In this particular case study, our simulation results and cost analysis suggest that the proposed transportation system in acceptable cost range considering the advantages that can be gained by using AMR system.

Recommendations for future work include taking into account the transport from workstations to storage, i.e. see how the replenishment affect performance of the proposed AMR system. Furthermore, battery management has to be considered as well. In this chapter, we assume that the robot is operational without the need of recharging the battery. In reality, the robots need to be recharged that results in lower throughput. Thus it is important to investigate how many additional robots are needed to keep the throughput the same. Lastly, Chap. 13 explains how detection of objects can be done using a camera system. It is worthwhile to integrate the system with the mobile robots so the robots have more advanced technique to detect unwanted or dangerous situations.

References

1. Andriansyah R, Etman LFP, Rooda JE (2010) Flow time prediction for a single-server order picking workstation using aggregate process times. Int J Adv Syst Meas 3:35–47
2. Arai T, Pagello E, Parker LE (2002) Guest editorial advances in multi-robot systems. IEEE Trans Robot Autom 18:655–661
3. Chen YQ, Wang Z (2005) Formation control: a review and a new consideration. In: 2005 IEEE/RSJ international conference on intelligent robots and systems, 2005 (IROS 2005), pp 3664–3669
4. ESI Falcon project (2011) Esifalcon's channel. http://www.youtube.com/user/ESIFalcon. Viewed May 2011
5. Gu J, Goetschalckx M, McGinnis LF (2010) Research on warehouse design and performance evaluation: a comprehensive review. Eur J Oper Res 203:539–549
6. Jiang ZP, Nijmeijer H (1997) Tracking control of mobile robots: a case study in backstepping. Automatica 33:1393–1399

7. Kanayama Y, Kimura Y, Miyazaki F, Noguchi T (1990) A stable tracking control method for an autonomous mobile robot. In: Proceedings., 1990 IEEE international conference on robotics and automation, pp 384–389
8. Kostić D, Adinandra S, Caarls J, Nijmeijer H (2010) Collision-free motion coordination of unicycle multi-agent systems. In: American control conference (ACC), pp 3186–3191
9. Kostić D, Adinandra S, Caarls J, van de Wouw N, Nijmeijer H (2009) Collision-free tracking control of unicycle mobile robots. In: Proceedings of the 48th IEEE Conference on decision and control, 2009 held jointly with the 2009 28th Chinese control conference, CDC/CCC 2009, pp 5667–5672
10. Lacomme P, Larabi M, Tcherne N (2010) Job-shop based framework for simultaneous scheduling of machines and automated guided vehicles. Int J Prod Econ (in press)
11. Liu S, Sun D, Zhu C (2010) Motion planning of multirobot formation. In: 2010 IEEE/RSJ international conference on intelligent robots and systems (IROS), pp 3848–3853
12. Mondada F, Bonani M, Raemy X, Pugh J, Cianci C, Klaptocz A, Magnenat S, Zufferey JC, Floreano D, Martinoli, A (2009) The e-puck, a robot designed for education in engineering. In: Proceedings of the 9th conference on autonomous robot systems and competitions, pp 59–65
13. Weyns D, Schelfthout K, Holvoet T, Lefever T (2005) Decentralized control of E'GV transportation systems. In: Proceedings of the fourth international joint conference on autonomous agents and multiagent systems, pp 67–74
14. Wurman PR, D'Andrea R, Mountz M (2008) Coordinating hundreds of cooperative, autonomous vehicles in warehouses. AI Mag 29:9–20

Part VI
Conclusion

Chapter 15
Reflections on the Falcon Project

Roelof Hamberg, Jacques Verriet, and Jan Schuddemat

Abstract This chapter reflects on Falcon, a project to advance automation in warehouses. Its main results and their impact as well as the project's process are discussed. The impact of Falcon on its industrial partners mainly concern model-based methods, of which strengthening the inception of the new architecture for system-level control of warehouses is a good example. Reflecting on Falcon as an industry-as-laboratory project, the application of obtained research results in an industrial context, albeit in an indirect way, has been perceived as successful. The project learnt us that the wide span of industrial challenges and research approaches in Falcon should be bridged in multiple steps with smaller, though overlapping scopes.

15.1 Introduction

The Falcon project started in 2006 with the common goal of providing the essential ingredients for the warehouse of the future. The project was defined as a BSIK industry-as-laboratory research project with Vanderlande Industries B.V., the Embedded Systems Institute, and six different groups from the three technical universities in the Netherlands as partners. An industry-as-laboratory research project [4] strives to obtain industrially applicable solutions to problems experienced in industry: as the prevailing research method it uses industry itself as a laboratory to validate the results. In 2011, Falcon concluded with three additional partners, i.e. Eurandom,

R. Hamberg (✉) · J. Verriet · J. Schuddemat
Embedded Systems Institute, P.O. Box 513, 5600 MB Eindhoven, The Netherlands
e-mail: roelof.hamberg@esi.nl

J. Verriet
e-mail: jacques.verriet@esi.nl

J. Schuddemat
e-mail: jan.schuddemat@esi.nl

R. Hamberg and J. Verriet (eds.), *Automation in Warehouse Development*,
DOI: 10.1007/978-0-85729-968-0_15, © Springer-Verlag London Limited 2012

Demcon Advanced Mechatronics, and Utrecht University, which entered the project after progressing insights in its goals.

This chapter gives a retrospective overview of the Falcon project. In Sect. 15.2, the most important results and their industrial impact are discussed. In Sect. 15.3, reflections on Falcon as an industry-as-laboratory project are given: the main specific activities and their observed effects are evaluated. Sections 15.4 and 15.5 summarise and conclude this chapter.

15.2 Overview of Technical Results and their Impact

This section combines a high-level overview of the research results obtained in the Falcon project with a discussion of their industrial impact. The reflection on these results and their impact is discussed per part of the book.

15.2.1 Decentralised Control Engineering

Part II of this book deals with the problem of arriving at a manageable, understandable, and yet flexible system-level control. The leading principle to achieve this was modularity. This principle appears a trivial starting point when one forgets about the scope and complexity of the systems to be controlled and the optimisations that are introduced to operate these systems. The latter situation refers to the observed state-of-practice in warehousing at the start of the project. Modularity requires distribution of control intent over multiple entities, which shifts the challenge for control from having oversight of the system to being able to engineer the resulting system behaviour. Several models have been made to support the latter with a specific focus on the modularisation of behaviour.

For architecting the system-level control of warehouses, it appeared that having multiple options improved the quality of decisions. The creation of an operational *holonic* control system [2, 3] was essential for putting system-level control on Falcon's agenda. The effort in researching decentralised control engineering was increased by adding a new partner to the project, Utrecht University. Although results were not immediately adopted in a literal sense, the mutual influence of the developments described in Chaps. 2 and 3 is very apparent in the interaction protocols, the division of responsibilities over different roles, and the locality of used information.

Modularity of control is a practical prerequisite for the application of model-based techniques. Part I of the book also focuses on the feasibility of such techniques to improve quality and efficiency. Quality improvement is obtained through a wider validation of behaviour upfront (e.g. by simulation), and safeguarded by using a more formal language to express the system behaviour. Efficiency gain is obtained through a warehouse domain-specific editor, controlling the start of a work-flow with automatic model transformations, which can cover generating framework code up to full control code including customisable component behaviours. Though being well-known in computer science research for some time, the model-driven approach, as discussed in Chap. 4, is new to the field of warehousing.

At the end of the project, the development of a warehouse domain-specific language is still ongoing. The model-based approach has been adopted by a team of developers in industry, who have built a prototype of a new system-level control in order to prove its feasibility and get feedback on the applied design patterns, interfaces, and business rules that were implemented to arrive at the required system behaviour. This prototype has delivered insights for a new platform for system-level controls in complex warehouse systems.

Naturally, the aforementioned challenges and results are part of a larger context in which many more problems are faced. Nevertheless, the Falcon project has provided an essential contribution for bootstrapping the new platform development within warehouse systems in a successful way.

15.2.2 Models in System Design

Nearly all research in the Falcon project is related to model-based design. Next to models serving as the starting point to engineer real products, such as software and hardware, models also play an important role in analysing systems before they actually exist. In Part III of this book, it has been discussed how simple models can be deployed to analyse system aspects in early phases of development. The main principle can be summarised by making models as simple as possible, but not too simple.

It has been long known that for warehousing the most critical system aspects include the system's throughput, timeliness (the ability to meet customer order deadlines), and availability. These aspects have in common that the involved component characteristics are not easily translated to the system level. Targeting at the simplification of the performance characteristics of relatively large subsystems, the application of the effective process time (EPT) method [1] in warehousing gave rise to some specific adaptations. These adaptations relate to switching times between orders and the problem of sequencing in warehousing, both of which are explained in Chap. 5. Although the quality of this research expressed in academic standards is good, the mismatch between the targeted industrial context (early-phase development of new systems) and the followed approach (measurements of existing systems to yield highly accurate predictions) prevented the results to be widely applied. Still, the development of complex warehouse systems can improve by more consciously applying the basic principle of EPT (i.e. model calibration through measurements at different aggregation levels).

In Part III, it is also illustrated that highly simplified models are feasible with a good cost/benefit ratio, in new warehouse concept development as well as in warehouse system configuration for specific customers. These models were perceived to have a large impact in the industrial context. One of the most visible examples is a black-box simulator that has been applied in several Vanderlande customer quotations. An essential success factor for such models to succeed is the low cost for applying them: the Falcon models are configured rather than made, which is done with well-known tools such as Microsoft Excel, and analyses can be made very fast. Through this, the overall feedback cycle of the modelling process becomes so short that many different variants can be investigated. This yields system insights.

The impact of these so-called analysis models is seen in the development of new concepts as well as in configuring specific customer systems from known components. Nevertheless, autonomous application of this method in industry is still a challenge. The creation of simple models compels to determine the essential characteristics of the problem, which is an important competence to develop in order to stay competitive. It is an advantageous perspective that system engineers and architects are able to create and use such models themselves in their daily work, especially when they are involved in the development of new concepts. It has been realised that the initial investment in learning the required methods is a blocking factor to achieve this. With Falcon, the awareness process with respect to these issues has started.

15.2.3 Automated Item Handling

In Part IV of this book, the feasibility of automated item picking has been studied. An automated solution for the item picking function is needed to arrive at the next level of efficient, flexible, and error-free warehousing. At the start of Falcon, the most difficult ingredients to arrive at a solution were identified as gripping many different items, appropriate sensing, and the coordination in between, resembling eye-hand coordination of humans.

From the outset, the research in gripping was inspired by bio-mechanical principles. The design of an automated gripper was done starting from the principles of underactuation for simple control and compliance for versatility, just like human hands work. At the end of the project, not only the feasibility of cost-effective grippers has been shown: a model-based design method to be able to redesign such grippers in other contexts has been developed in addition to that as well.

The application of computer vision to sense and control item picking faced a specific challenge: selecting the optimal item to pick from a scene with mainly partial occluded identical objects is not trivial. A smart match between learning the salient features of a product type once and combining 2D and 2.5D sensory information for every pick action in a fast and robust way was shown to tackle the selection challenge effectively.

The initial research taken up in gripper design and computer vision was parallelled by the development of an automated item-picking system setup by Vanderlande. The latter development took a much more pragmatic approach which resulted in some choices that were faster but less versatile than the available research results. Parts of the computer vision algorithms were taken on board, whereas the gripping hands were replaced by vacuum suction cups. The two tracks of development in the project dispersed further with Vanderlande's adequate initiative of the automated item-picking workstation. The basic technology elements were made even simpler, and the focus was directed towards suitable integration of such a workstation in a complete warehouse system in an evolutionary way. Ultimately, the range of products that can be handled by this workstation is too small for a complete system solution, but it still sets an important benchmark to be integrated in hybrid system solutions.

Ergo, largely inspired by the advanced research undertaken in Falcon, Vanderlande prototyped an automated item-picking workstation to show the feasibility of a solution built from components available in the marketplace. To match this with possible applications of new research results an additional industrial partner was found. This was Demcon Advanced Mechatronics, who entered the project in 2008.

Allegedly, the integration of new gripper designs and computer vision algorithms have increased the business opportunities of Demcon. This is mainly achieved through increased competence in the aforementioned key technologies, which improved the market position of Demcon. Ultimately, the versatile gripper design and the computer vision algorithms developed by the research partners were shown to outperform the benchmark provided by the system of Vanderlande.

In the course of the Falcon project, Vanderlande also started the development of an automated case picking system. Case picking is simpler than item picking, and such systems have good market potential. The developed system is remarkably in concordance with the earlier developed automated item-picking system concept developed in Falcon. Although the latter one has never arrived at any level of practice, it did serve as an initial point of reference for Vanderlande and that should be noticed as one of the points of impact of the Falcon project.

15.2.4 Transport by Roaming Vehicles

Part V of this book has been devoted to flexible transport solutions in warehousing. The well-known conveyor-based solutions are proven technology: one should be cautious to change them. Nevertheless, they also have disadvantages: conveyors are far from flexible after installation and therefore not very suited to support changing customer business processes. It is a challenge to look for possibilities to deal with the problems without endangering the established successes.

The challenge of realising the transport function in a more flexible way has been met by the Falcon project. The research partners expected that performing transport with autonomous vehicles that roam around would provide an alternative, at the cost of introducing new challenges of self-localisation and mutual coordination. Although the challenges are manifold for this new technology, the associated benefits are also very large. The versatility of vehicles can be applied, not only for having a more continuous and therefore better scalability in terms of throughput capacity, but also for obtaining a much more robust system as shuttle failures have only a graceful degradation effect on the total system availability (provided the right provisions have been taken).

In the last part of this book, the critical aspects of a feasible and economic alternative to the well-known conveyor technology have been dealt with. It has been illustrated how high-level and low-level coordination algorithms can cooperate to deliver the necessary coordination between several vehicles. The stability and performance of decentralised control are shown. It has been shown that computer vision can be applied to sense and control in a robust way in situations where self-localisation is needed in an incompletely structured and calibrated environment that will occasionally exhibit unexpected situations.

It is a challenge to align the time horizons of the long-term research results and the industrial need to deliver working products to the marketplace. The momentum of proven conveyor technology is still determining the way projects are being realised. Nevertheless, the issue of roaming shuttles is more frequently on the agenda within Vanderlande, which fact is inalienably related to the research activities in Falcon.

15.3 New Elements for Industry-as-Laboratory Projects

In this section, new activities and events of Falcon in its context of being an industry-as-laboratory project [4] are discussed.

15.3.1 Setting the Broader Context

When embarking on an industry-as-laboratory project, one of the most challenging factors is getting to know the domain of the industry that provides the project's problem statement. A number of activities have been conducted to achieve this in Falcon.

First, a series of so-called knowledge transfer sessions was organised, comparable to in-house courses that many industries organise for new employees. Representatives of Vanderlande introduced their department and way of working. These included sales, simulation, reference architecture, research and development, engineering, software, and service. The knowledge transfer sessions served multiple goals: not only are they an effective and efficient way to get the project's freshmen organisational insights at a necessary base level, they also provide a common set of insights whereby the communication about Vanderlande's domain has become more natural. It is recommendable to share such sessions with the whole project team.

Second, several visits have been conducted to customers of Vanderlande in order to observe operational warehouse systems. For our context, warehousing, the wide variety of handled goods and the scale of different streams (such as the waste flow) do become very distinct through such visits. Remarks of customer employees help to judge the relative importance of system requirements. One customer visit in the UK has been important because the human item picker's job (candidate for automation) has been carried out for a few days. This yielded information about the constituent tasks as well as about item classifications. Another customer visit has been very relevant for the system modelling research: the EPT approach has been validated with real-life data from that system. This immersed the researchers in the practical situation of dealing with non-ideal data.

Third, the academic groups were visited with the whole project team in order to share knowledge about the context the researchers are working in. This further strengthened the common frame of reference in a bidirectional way.

Concluding, these context-setting activities have been valuable in order to get a common understanding of the industrial domain, industrial problems, the project goals, the possibilities to apply research results, and to build the Falcon team. The feedback from these activities is relevant for the whole project team, as the investment in understanding context is important for solving the right problems.

15.3.2 Architecture Team and Reference Cases

The Falcon project started with the common goal of providing the essential ingredients for the warehouse of the future. A number of activities were pursued to translate and couple this goal to research directions of the contributing partners in the project.

First, the complete project team has been involved in brainstorm sessions in order to come up with concepts for this warehouse of the future as well as its essential ingredients. The gap between the industrial challenge and the research themes of the groups appeared to be too large.

Second, the so-called *Warehouse Architecture Team* (WHAT) was installed. This mixed team has been working on architectural views for the warehouse of the future, based on a reference customer case. Their work started from the prevailing reference architecture in Vanderlande, discussing its problematic issues. However, due to lack of conclusive decisions on how to link the industrial problems to research questions, the connection of their results to practical research goals failed.

Third, the demonstrator days (see also Sect. 15.3.3) were programmed along the lines of a global vision including a new system concept. A larger degree of connection was achieved, but still the span of topics of the Falcon project was too large to create sensible overlaps between all parts of research. A second reference customer case was defined, including extensive data analyses, but hardly used by researchers in practice. Nevertheless, the system concept has served as inspiration for new developments within Vanderlande.

Concluding, the aforementioned activities have to be evaluated as good attempts to provide the Falcon researchers with a common framework. The balancing act between the two layers, industry and academia, has been largely unproductive. The compliant recommendation for industry-as-laboratory projects would be to sincerely consider whether or not the different project partners' objectives can be bridged in one single step. In hindsight this has not been possible in Falcon.

15.3.3 Demonstrator Days

The Falcon project team has organised two demonstrator days in order to communicate research plans, approaches, and results to Vanderlande. The definition of demonstrators for such days has the immediate result of making goals much more concrete. Among the researchers an increased level of urgency is perceived for reaching those goals, which however are not always felt as entirely their own goals. The same holds for creating the demonstrators, which required executing some non-core (i.e. non-research) activities. Nevertheless, these activities are considered essential for effective transfer of results to the industrial partners, which not only depends on the quality of those results, but also on the industrial acceptance level that the proposed solutions will work. Demonstrators should be part of the research, a vehicle to get feedback, a first step in validation.

The demonstrator days themselves provide ample opportunities to find more relationships between research and parts of industry. Nevertheless, it is mainly a matter of getting acquainted to each other, while the follow-up of such contacts has proved to be difficult and needs focus and effort. Some of the topics have been presented in the form of a workshop. The advantage of the workshop format is that more response is generated, not only because the setting is very interactive, but also due to participants reacting on each other. Still, also here a follow-up does not come for free: the workshop itself can only be the start of a process.

15.3.4 Partner Changes

Sizable, multi-annual projects are inevitably subject to changes. The probability of changes is so large that one better prepares for this from the start.

Remarkable changes in Falcon (compared to other industry-as-laboratory projects of ESI) were the involvement of a second so-called carrying industrial partner, Demcon Advanced Mechatronics, and an additional academic partner, Utrecht University. The first change was employed in order to strengthen the match between a valid industrial context and the research conducted in the area of mechanical design and computer vision (see Sect. 15.2.3). The second change was a measure to increase the amount of research effort into system-level control approaches (see Sect. 15.2.1).

Reflecting on these changes, a few more conclusions can be drawn. First, the urgency of the challenges in the industrial context should be very clear for the involved researchers. This requires not only their topics to match, but it also appeared that an industrial roadmap is key to conduct focused and effective research. Second, the initial problem statement of a project like Falcon needs significant effort to become more clear. Albeit difficult to realise in practice, starting with a small team (much alike WHAT) to better define the challenge, and only scale up the project team when the problem statement has become very crisp, is a better approach of conducting industry-as-laboratory projects. The more detailed problem statement enables to hypothesise solution directions for which academic partners can be found to work on.

Key persons are also bound to change jobs during the lifetime of the project. Ownership roles change, and the probability for this is largest at the industrial side. In the course of Falcon, almost every manager of the involved departments of Vanderlande (research & development, systems engineering, simulation) has changed his role, but also in academia and at the Embedded Systems Institute such changes have occurred albeit to a lesser degree. These changes require continuous adaptation of the project team. The evolving nature of the project (from understanding the industrial domain at the start to having results and impact at the end) is also reflected in changes of the network of key persons at the industrial side. Nevertheless, the impact of Falcon can largely be attributed to a small number of industrial people who have played a stable role throughout the project's lifetime.

15.3.5 Embedding in Industry

A practical method to influence the nature of the research being conducted is to change the balance of *where* the research is being conducted. It appears that the sheer location where a researcher is working has a significant impact on at least the packaging of results, but often also on the unfurled activities themselves.

This fact has been observed with students that have been situated at Vanderlande Industries and Demcon Advanced Mechatronics. In general, they have delivered more practical results than students without any home in industry, a fact which is highly recognised by the industrial partners. For PhD students, this gives rise to a tension with the fact that they have to deliver strictly new research results in scientific publications. In general, the combination of obtaining practical results for industry and reflecting on those results scientifically was only observed with very good students.

In the latter half of the Falcon project, the research fellows of ESI have spent a significant amount of their time at the main site of Vanderlande. This enabled them in a very natural way to better weigh the different research directions and questions by monitoring the development teams in industry. Such improved observations deliver more direct feedback on what does really work in practice and what does not. Finally, this way of working had a much greater impact on the daily activities as performed in industry and should have been started much earlier in the project.

15.4 Summary

The aim of the Falcon project was to advance automation in warehousing. Its main results with industrial impact are:

- Development of a new modular approach for distributed control including constituent behaviours in order to implement system-level control in a manageable, understandable, and yet flexible way.

- Application of the domain-specific language approach in warehousing to improve quality and efficiency.
- Feasibility and application of early-phase dynamic models of warehouse systems to support new concept development and system configuration.
- Design and design method for item grippers that outperform any off-the-shelf solution in terms of versatility.
- Development of a computer vision system to learn to recognise items that outperforms any off-the-shelf solution in terms of versatility.
- Development of an integrated item picking setup that outperforms any other known alternative solution in terms of versatility.
- Feasibility of controlling shuttle-based transport to increase flexibility.
- Feasibility of a computer vision system to support simultaneous self-localisation and map building in a semi-structured environment.

The main recommendations for conducting industry-as-laboratory projects are:

- Invest in sharing the industrial context and problems with the whole project team.
- Bridging gaps between industrial problems and research approaches might require more than one step. Investigate mutual roadmaps and mindsets.
- Define specific customer cases to make the system context and usage clear.
- Transfer of results requires focus and endurance of both providers and receivers. Events, such as demonstrator days, are catalysts, but require active follow-up.
- Bootstrap projects in phases: start concise and expand later.
- Double-check commitment with all involved partners, especially after role and person changes. Pay sufficient attention to the associated managerial change process.
- Conduct research at the location where its results should be applied for at least three days per week.

15.5 Concluding Remarks

This chapter concludes the Falcon book, a book that summarises the achievements of the Falcon project. In 2006 the project team in the making took up the challenge to automate item picking in the context of the warehouse of the future. The partners united under the colours of "industry-as-laboratory" philosophy to embark on a research journey with many ups and downs.

The warehouse of the future is not there, and should it come, a new horizon in warehousing is always to be discovered. Nevertheless, Falcon has brought several exciting ingredients not to be scorned in a new generation of Flexible Automated Logistics CONcepts (FALCON).

References

1. Jacobs JH, Etman LFP, van Campen EJJ, Rooda JE (2003) Characterization of operational time variability using effective processing times. IEEE Trans Semicond Manuf 16:511–520
2. Moneva H, Caarls J, Verriet J (2009) A holonic approach to warehouse control. In: 7th international conference on practical applications of agents and multi-agent systems (PAAMS 2009). Advances in intelligent and soft computing, vol 55. Springer, Heidelberg, pp 1–10
3. Moneva HG (2008) A holonic approach to decentralized warehouse control. SAI technical report, Eindhoven University of Technology, Eindhoven
4. Potts C (1993) Software-engineering research revisited. IEEE Softw 10:19–28

Appendix A
Falcon Project Partners

This appendix provides an overview of the organisations that have participated in the Falcon project.

- Embedded Systems Institute, P.O. Box 513, 5600 MB Eindhoven, The Netherlands
- Vanderlande Industries B.V., Vanderlandelaan 2, 5466 RB Veghel, The Netherlands
- Demcon Advanced Mechatronics, Zutphenstraat 25, 7575 EJ Oldenzaal, The Netherlands
- Delft University of Technology, Faculty of Mechanical, Maritime and Materials Engineering (3ME), Department BioMechanical Engineering, P.O. Box 5, 2600 AA Delft, The Netherlands
- Eindhoven University of Technology, Department of Mathematics and Computer Science, Software Engineering and Technology Group, P.O. Box 513, 5600 MB Eindhoven, The Netherlands
- Eindhoven University of Technology, Department of Mathematics and Computer Science, Stochastic Operations Research Group, P.O. Box 513, 5600 MB Eindhoven, The Netherlands
- Eindhoven University of Technology, Department of Mechanical Engineering, Dynamics and Control Group, P.O. Box 513, 5600 MB Eindhoven, The Netherlands
- Eindhoven University of Technology, Department of Mechanical Engineering, Manufacturing Networks Group, P.O. Box 513, 5600 MB Eindhoven, The Netherlands
- University of Twente, Faculty of Electrical Engineering, Mathematics and Computer Science, Control Engineering Group, P.O. Box 217, 7500 AE Enschede, The Netherlands
- Utrecht University, Faculty of Science, Department of Information and Computing Sciences, Decision Support Group, P.O. Box 80.089, 3508 TB Utrecht, The Netherlands
- Eurandom, P.O. Box 513, 5600 MB Eindhoven, The Netherlands

R. Hamberg and J. Verriet (eds.), *Automation in Warehouse Development*,
DOI: 10.1007/978-0-85729-968-0,

Appendix B
Falcon Project Publications

This appendix provides an overview of the articles, papers, reports, and theses published within the scope of the Falcon project.

1. Adinandra S, Caarls J, Kostić D, Nijmeijer H (2010) Performance of high-level and low-level control for coordination of mobile robots. In: Proceedings of the 7th international conference on informatics in control, Automation and Robotics (ICINCO), pp 63–71
2. Adinandra S, Kostić D, Caarls J, Nijmeijer H (2011) Towards a flexible and scalable transportation in distribution centers: low-level motion control approach. In: Proceedings of the 8th international conference on informatics in control, Automation and Robotics (ICINCO), pp 155–160
3. Aertssen J, Rudinac M, Jonker P (2011) Fall and action detection in elderly homes. In: Conference on advancement of assistive technology in Europe (AAATE)
4. Aertssen J, Rudinac M, Jonker P (2011) Real time fall detection and pose recognition in home environments. In: international joint conference on computer vision, Imaging and computer graphics theory and applications (VISAPP)
5. Akman O (2011) Detection, tracking and mapping for mobile robots and augmented reality in context-free environments. Ph.D. thesis, Delft University of Technology, Delft
6. Akman O, Bayramoglu N, Alatan AA, Jonker P (2010) Utilization of spatial information for point cloud segmentation. In: 3DTV-conference: the true vision—capture, Transmission and display of 3D Video (3DTV-CON), pp 1–4
7. Akman O, Jonker P (2009) Exploitation of 3d information for directing visual attention and object recognition. In: Proceedings of the eleventh IAPR conference on machine vision applications, pp 50–53
8. Akman O, Jonker P (2010) Computing saliency map from spatial information in point cloud data. In: Advanced concepts for intelligent vision systems, Lecture notes in computer science, vol 6474. Springer, Berlin, pp 290–299
9. Akman O, Lenseigne B, Jonker P (2009) Directing visual attention and object recognition using 3d information. In: Proceedings of the fifteenth annual conference of the advanced school for computing and imaging

R. Hamberg and J. Verriet (eds.), *Automation in Warehouse Development*,
DOI: 10.1007/978-0-85729-968-0,

10. Aldewereld H, Dignum F, Hiel M (2011) Reorganization in warehouse management systems. In: Proceedings of the IJCAI 2011 workshop on artificial intelligence and logistics (AILog-2011), pp 67–72
11. Andriansyah R (2011) Order-picking workstations for automated warehouses. Ph.D. thesis, Eindhoven University of Technology, Eindhoven
12. Andriansyah R, Etman LFP, Adan IJBF, Rooda JE (2011) Automated order-picking workstation handling out-of-sequence product arrivals. In: Proceedings of the 1st international conference on simulation and modeling methodologies, Technologies and applications, pp 283–292
13. Andriansyah R, Etman LFP, Rooda JE (2009) On sustainable operation of warehouse order picking systems. In: XIV Summer School 'Francesco Turco', pp IV.16–IV.23
14. Andriansyah R, Etman LFP, Rooda JE (2009) Simulation model of a single-server order picking workstation using aggregate process times. In: Advances in system simulation, 2009. SIMUL '09. First international conference on, pp 23–31
15. Andriansyah R, Etman LFP, Rooda JE (2010) Aggregate modeling for flow time prediction of an end-of-aisle order picking workstation with overtaking. In: Winter simulation conference (WSC), Proceedings of the 2010, pp 2070–2081
16. Andriansyah R, Etman LFP, Rooda JE (2010) Flow time prediction for a single-server order picking workstation using aggregate process times. Int J Adv Syst Meas 3:35–47
17. Andriansyah R, de Koning WWH, Jordan RME, Etman LFP, Rooda JE (2008) Simulation study of miniload-workstation order picking system. SE-Report 2008-07, Eindhoven University of Technology, Department of Mechanical Engineering, Eindhoven
18. Andriansyah R, de Koning WWH, Jordan RME, Etman LFP, Rooda JE (2011) A process algebra based simulation model of a miniload workstation order picking system. Comput Ind 62:292–300
19. Ansems RPWM (2008) Scheduling the unloading of incoming containers in a distribution center. Bachelor's thesis, Eindhoven University of Technology, Department of Mechanical Engineering, Systems Engineering Group, Eindhoven
20. Arnoldus J, Bijpost J, van den Brand M (2007) Repleo: a syntax-safe template engine. In: Proceedings of the 6th international conference on generative programming and component engineering, pp 25–32
21. Baril M (2011) Stable precision grasps with underactuated fingers. Internship report, Delft University of Technology, Department of BioMechanical Engineering, Delft
22. Bayramoglu N, Akman O, Alatan AA, Jonker P (2009) Integration of 2d images and range data for object segmentation and recognition. In: Proceedings of the twelfth international conference on climbing and walking robots and the support technologies for mobile machines, pp 927–933
23. Berendse DFJ (2010) Design, verification and analysis of the highly dynamic storage system. Master's thesis, Eindhoven University of Technology, Department of Mathematics and Computer Science, Eindhoven

24. Bijl RJ (2010) Formalizing material flow diagrams. Master's thesis, Eindhoven University of Technology, Department of Mathematics and Computer Science, Eindhoven
25. Birglen L, Kragten GA, Herder JL (2010) State-of-the-art in underactuated grasping. Mech Sci 1:3
26. Bos HD (2010) Evolution of robotic hands. Internship report, University of Twente, Control Laboratory, Enschede
27. Bosch A, Slobbe J, van Dam T (2008) Het effect van contactmateriaal van een robothand. Bachelor's thesis, Delft University of Technology, Department of BioMechanical Engineering, Delft
28. Bouarfa L, Akman O, Schneider A, Jonker PP, Dankelman J (2011) In-vivo real-time tracking of surgical instruments in endoscopic video. Minim Invasive Ther Allied Technol
29. Chang Y (2010) Design of an underactuated gripper for the item picking in distribution centers. Master's thesis, Delft University of Technology, Department of BioMechanical Engineering, Delft
30. De Jong D (2007) Analyse naar het grijpen met een Soft Gripper: een vergelijking tussen praktijk en computersimulatie. Bachelor's thesis, Delft University of Technology, Department of BioMechanical Engineering, Delft
31. De Koning WWH (2008) Modeling a storage and retrieval system: architecture and model aggregations. Master's thesis, Eindhoven University of Technology, Department of Mechanical Engineering, Systems Engineering Group, Eindhoven
32. De Natris R (2010) Analyse van een AIP station met een eindige buffer. Bachelor's thesis, Eindhoven University of Technology, Department of Mechanical Engineering, Systems Engineering Group, Eindhoven
33. Den Dunnen S (2009) The design of an adaptive finger mechanism for a hand prosthesis. Master's thesis, Delft University of Technology, Department of BioMechanical Engineering, Delft
34. Differ HG (2010) Design and implementation of an impedance controller for prosthetic grasping. Master's thesis, University of Twente, Control Laboratory, Enschede
35. Differ HG (2010) Development of a homing procedure and investigation of tip stiffness for the robotic finger test setup. Internship report, University of Twente, Control Laboratory, Enschede
36. Engelen L, van den Brand M (2010) Integrating textual and graphical modelling languages. Electron Notes Theor Comput Sci 253:105–120
37. Febrianie B (2011) Queueing models for compact picking systems. Master's thesis, Eindhoven University of Technology, Department of Mathematics and Computer Science, Eindhoven
38. Ficuciello F, Carloni R, Visser LC, Stramigioli S (2010) Port-Hamiltonian modeling for soft-finger manipulation. In: Intelligent Robots and Systems (IROS), 2010 IEEE/RSJ international conference on, pp 4281–4286

39. Giaccotto R (2008) Smooth surface fitting by patches, a new method of interpolation for contact modeling. Master's thesis, University of Twente, Control Laboratory, Enschede
40. Guitian Mediero PJ (2009) Modular platform for the experimental evaluation of underactuated finger. Internship report, Delft University of Technology, Department of BioMechanical Engineering, Delft
41. Hakobyan L (2009) Warehouse design toolbox. SAI technical report, Eindhoven University of Technology, Eindhoven
42. Hamberg R (2008) Tilt-tray sorters modelled with UPPAAL. ESI Report 2008-2, Embedded System Institute, Eindhoven
43. Heling JWE (2011) Design of an automated item picking workstation. Master's thesis, Eindhoven University of Technology, Department of Mechanical Engineering, Manufacturing Networks Group, Eindhoven
44. Hiel M, Aldewereld H, Dignum F (2010) Modeling warehouse logistics using agent organizations. In: Collaborative agents—research and development, Lecture notes in computer science, vol 6066. Springer, Berlin, pp 14–30
45. Jordan RME (2007) Literature review on designing a warehouse order picking system with conveyors and workstations. Internship report, Eindhoven University of Technology, Department of Mechanical Engineering, Systems Engineering Group, Eindhoven
46. Jordan RME (2008) Modeling the item picking area of the plus retail compact picking system. Master's thesis, Eindhoven University of Technology, Department of Mechanical Engineering, Systems Engineering Group, Eindhoven
47. Kavuma JM (2009) Holonic highly dynamic storage system: agent-based distributed control. SAI technical report, Eindhoven University of Technology, Eindhoven
48. Kool AC (2008) Grasping performance in compliant underactuated robotic hands. Master's thesis, Delft University of Technology, Department of BioMechanical Engineering, Delft
49. Kostić D, Adinandra S, Caarls J, Nijmeijer H (2010) Collision-free motion coordination of unicycle multi-agent systems. In: American control conference (ACC), 2010, pp 3186–3191
50. Kostić D, Adinandra S, Caarls J, van de Wouw N, Nijmeijer H (2009) Collision-free tracking control of unicycle mobile robots. In: Decision and control, 2009 held jointly with the 2009 28th Chinese control conference. CDC/CCC 2009. Proceedings of the 48th IEEE conference on, pp 5667–5672
51. Kostić D, Adinandra S, Caarls J, van de Wouw N, Nijmeijer H (2010) Saturated control of time-varying formations and trajectory tracking for unicycle multi-agent systems. In: Decision and control (CDC), 2010 49th IEEE conference on, pp 4054–4059
52. Kragten GA (2011) Underactuated hands: fundamentals, performance analysis and design. Ph.D. thesis, Delft University of Technology, Delft
53. Kragten GA, Baril M, Gosselin C, Herder JL (2011) Stable precision grasps by underactuated fingers. IEEE Trans Rob, To appear

54. Kragten GA, Bosch HA, van Dam T, Slobbe JA, Herder JL (2009) On the effect of contact friction and contact compliance on the grasp performance of underactuated hands. In: ASME 2009 international design engineering technical conferences and computers and information in engineering conference, vol 7, pp 871–878
55. Kragten GA, van der Helm FCT, Herder JL (2011) A planar geometric design approach for a large grasp range in underactuated hands. Mech Mach Theory 46:1121–1136
56. Kragten GA, Herder JL (2007) Equilibrium, stability and robustness in underactuated grasping. In: ASME 2007 international design engineering technical conferences and computers and information in engineering conference, vol. 8, pp 645–652
57. Kragten GA, Herder JL (2010) The ability of underactuated hands to grasp and hold objects. Mech Mach Theory 45:408–425
58. Kragten GA, Herder JL (2010) A platform for grasp performance assessment in compliant or underactuated hands. J Mech Des 132:1–6
59. Kragten GA, Herder JL, Schwab AL (2008) On the influence of contact geometry on grasp stability. In: ASME 2008 international design engineering technical conferences and computers and information in engineering conference, vol. 2, pp 993–998
60. Kragten GA, Kool AC, Herder JL (2009) Ability to hold grasped objects by underactuated hands: performance prediction and experiments. In: Robotics and Automation, 2009. ICRA '09. IEEE international conference on, pp 2493–2498
61. Kragten GA, Meijneke C, Herder JL (2010) A proposal for benchmark tests for underactuated or compliant hands. Mech Sci 1:13–18
62. Lassooij J, Reuijl D, Steenbergen R, Warnar P (2008) Ondergeactueerde robothand: van model naar ontwerp. Bachelor's thesis, Delft University of Technology, Department of BioMechanical Engineering, Delft
63. Liang HL (2011) A graphical specification tool for decentralized warehouse control systems. SAI technical report, Eindhoven University of Technology, Eindhoven
64. Liu L, Adan IJBF (2011) Queueing network analysis of compact picking systems. Working paper
65. Meijneke C, Kragten GA, Wisse M (2011) Design and performance assessment of an underactuated hand for industrial applications. Mech Sci 1:9–15
66. Mennink T (2010) Virtualization of the FALCON humanoid finger into a direct drive system. Internship report, University of Twente, Control Laboratory, Enschede
67. Meulen MG (2010) Verification of PLC source code using propositional logic. Master's thesis, Eindhoven University of Technology, Department of Mathematics and Computer Science, Eindhoven
68. Moneva H, Caarls J, Verriet J (2009) A holonic approach to warehouse control. In: 7th international conference on practical applications of agents and multi-agent systems (PAAMS 2009), Advances in intelligent and soft computing, vol 55. Springer, Berlin, pp 1–10

69. Moneva HG (2008) A holonic approach to decentralized warehouse control. SAI technical report, Eindhoven University of Technology, Eindhoven
70. Nguyen PH (2010) Quantitative analysis of model transformations. Master's thesis, Eindhoven University of Technology, Department of Mathematics and Computer Science, Eindhoven
71. Ordóñez Camacho D, Mens K, van den Brand M, Vinju J (2010) Automated generation of program translation and verification tools using annotated grammars. Sci Comput Program 75:3–20
72. Ouwerkerk B, Crooijmans B, de Nooij M, de Vries S (2008) Grijpbereik van een vormadaptieve robothand bij verschillende stijfheidverhoudingen tussen de kootjes. Bachelor's thesis, Delft University of Technology, Department of BioMechanical Engineering, Delft
73. Paese M (2008) Analysis of an automated item picking workstation. Master's thesis, Eindhoven University of Technology, Department of Mechanical Engineering, Systems Engineering Group, Eindhoven
74. Pape R (2011) The effect of joint locks in underactuated hand prostheses. Master's thesis, Delft University of Technology, Department of BioMechanical Engineering, Delft
75. Protić Z (2011) Configuration management for models: generic models for model comparison and model co-evolution. Ph.D. thesis, Eindhoven University of Technology, Eindhoven
76. Pulcini G (2010) Design of a miniaturized joint lock for an under actuated robotic finger. Master's thesis, University of Twente, Control Laboratory, Enschede
77. Reehuis E, Bäck T (2010) Mixed-integer evolution strategy using multiobjective selection applied to warehouse design optimization. In: Proceedings of the 12th annual conference on genetic and evolutionary computation, pp 1187–1194
78. Roode V (2011) Exception handling in automated case picking. SAI technical report, Eindhoven University of Technology, Eindhoven
79. Rudinac M, Jonker PP (2009) Entropy based method for keypoint selection. In: Proceedings of the fifteenth annual conference of the advanced school for computing and imaging
80. Rudinac M, Jonker PP (2010) A fast and robust descriptor for multiple-view object recognition. In: Control automation robotics & vision (ICARCV), 2010 11th international conference on, pp 2166–2171
81. Rudinac M, Jonker PP (2010) How to focus robots attention? In: Intelligent machines symposium
82. Rudinac M, Jonker PP (2010) Saliency based method for object localization. In: Proceedings of the sixteenth annual conference of the advanced school for computing and imaging
83. Rudinac M, Jonker PP (2010) Saliency detection and object localization in indoor environments. In: Pattern recognition (ICPR), 2010 20th international conference on, pp 404–407

84. Rudinac M, Jonker PP (2010) Scene exploration and object inspection for mobile robots in indoor environments. In: Bits&Chips embedded systemen symposium
85. Rudinac M, Jonker PP (2011) Visual categorization of unknown objects for mobile robotic applications. In: Hightech mechatronica
86. Rudinac M, Lenseigne B, Jonker P (2009) Keypoints extraction and selection for recognition. In: Proceedings of the eleventh IAPR conference on machine vision applications
87. Stel DWJ (2011) The impact of sequence requirements of product totes on the performance of a goods-to-man system. Master's thesis, Eindhoven University of Technology, Department of Mechanical Engineering, Manufacturing Networks Group, Eindhoven
88. Steutel P (2009) Design of a fully compliant underactuated finger with a monolithic structure and distributed compliance. Master's thesis, Delft University of Technology, Department of BioMechanical Engineering, Delft
89. Steutel P, Kragten GA, Herder JL (2010) Design of an underactuated finger with a monolithic structure and largely distributed compliance. In: ASME 2010 international design engineering technical conferences and computers and information in engineering conference, vol. 2, pp 355–363
90. Sun T (2010) Comparison and improvements of compact picking system models. Master's thesis, Eindhoven University of Technology, Department of Mathematics and Computer Science, Eindhoven
91. Van Amstel MF (2010) The right tool for the right job: assessing model transformation quality. In: Computer software and applications conference workshops (COMPSACW), 2010 IEEE 34th annual, pp 69–74
92. Van Amstel MF (2012) Assessing and improving the quality of model transformations. Ph.D. thesis, Eindhoven University of Technology, Eindhoven
93. Van Amstel MF, Bosems S, Kurtev I, Ferreira Pires L (2011) Performance in model transformations: a comparison between ATL and QVT. In: Theory and practice of model transformations: proceedings of the fourth international conference on model transformation (ICMT 2011), Lecture notes in computer science, vol 6707. Springer, Berlin, pp 198–212
94. Van Amstel MF, van den Brand MGJ (2010) Quality assessment of ATL model transformations using metrics. In: Proceedings of the second international workshop on model transformation with ATL (MtATL 2010)
95. Van Amstel MF, van den Brand MGJ (2011) Model transformation analysis: staying ahead of the maintenance nightmare. In: Theory and practice of model transformations: proceedings of the fourth international conference on model transformation (ICMT 2011), Lecture notes in computer science, vol 6707. Springer, Berlin, pp 108–122
96. Van Amstel MF, van den Brand MGJ (2011) Using metrics for assessing the quality of ATL model transformations. In: Proceedings of the third workshop on model transformations with ATL (MtATL2011), pp 20–34

97. Van Amstel MF, van den Brand MGJ, Engelen, LJP (2010) An exercise in iterative domain-specific language design. In: Proceedings of the joint ERCIM workshop on software evolution (EVOL) and international workshop on principles of software evolution (IWPSE), pp 48–57
98. Van Amstel MF, van den Brand MGJ, Engelen LJP (2011) Using a DSL and fine-grained model transformations to explore the boundaries of model verification. CS-report 11-02, Eindhoven University of Technology, Department of Mathematics and Computer Science, Eindhoven
99. Van Amstel MF, van den Brand MGJ, Engelen LJP (2011) Using a DSL and fine-grained model transformations to explore the boundaries of model verification. In: Proceedings of the third workshop on model-based verification & validation from research to practice (MVV 2011)
100. Van Amstel MF, van den Brand MGJ, Engelen LJP (2011) Using a DSL and fine-grained model transformations to explore the boundaries of model verification—extended abstract. In: Proceedings of the seventh workshop on advances in model based testing (A-MOST 2011)
101. Van Amstel MF, van den Brand MGJ, Nguyen PH (2010) Metrics for model transformations. In: Proceedings of the ninth Belgian-Netherlands software evolution workshop (BENEVOL 2010)
102. Van Amstel MF, van den Brand MGJ, Protić Z (2008) Version control of graphs. In: Informal pre-proceedings of the seventh Belgian-Netherlands software evolution workshop (BENEVOL 2008), pp 11–12
103. Van Amstel MF, van den Brand MGJ, Protić Z, Verhoeff T (2008) Transforming process algebra models into UML state machines: bridging a semantic gap? In: theory and practice of model transformations (Lecture notes in computer science), vol 5063. Springer, Berlin, pp 61–75
104. Van Amstel MF, Lange CFJ, van den Brand MGJ (2008) Metrics for analyzing the quality of model transformations. In: Proceedings of the twelfth ECOOP workshop on quantitative approaches on object oriented software engineering (QAOOSE 2008), pp 41–51
105. Van Amstel MF, Lange CFJ, van den Brand MGJ (2008) Metrics for analyzing the quality of model transformations—extended abstract. In: Informal pre-proceedings of the seventh Belgian-Netherlands software evolution workshop (BENEVOL 2008), pp 36–37
106. Van Amstel MF, Lange CFJ, van den Brand MGJ (2009) Using metrics for assessing the quality of ASF+SDF model transformations. In: Theory and practice of model transformations, Lecture notes in computer science, vol 5563. Springer, Berlin, pp 239–248
107. Van Amstel MF, van de Plassche E, Hamberg R, van den Brand MGJ, Rooda JE (2007) Performance analysis of a palletizing system. SE-report 2007-09, Eindhoven University of Technology, Department of Mechanical Engineering, Eindhoven
108. Van den Brand M, Protić Z, Verhoeff T (2010) Fine-grained metamodel-assisted model comparison. In: Proceedings of the 1st international workshop on model comparison in practice, pp 11–20

109. Van den Brand M, Protić Z, Verhoeff T (2010) Generic tool for visualization of model differences. In: Proceedings of the 1st international workshop on model comparison in practice, pp 66–75
110. Van den Brand M, Protić Z, Verhoeff T (2010) RCVDiff—a stand-alone tool for representation, calculation and visualization of model differences. In: Proceedings of international workshop on models and evolution—ME 2010
111. Van den Brand M, Protić Z, Verhoeff T (2011) A generic solution for syntax-driven model co-evolution. In: Proceedings of the 49th international conference on objects, models, components, patterns
112. Van den Brand MGJ, van der Meer AP, Serebrenik A (2009) Type checking evolving languages with MSOS. In: Semantics and algebraic specification, Lecture notes in computer science, vol 5700. Springer, Berlin pp 207–226
113. Van den Brand MGJ, van der Meer AP, Serebrenik A, Hofkamp AT (2010) Formally specified type checkers for domain specific languages: experience report. In: Proceedings of the tenth workshop on language descriptions, Tools and Applications, pp 12:1–12:7
114. Van den Brandt M (2010) US technological innovation systems for service robotics. Master's thesis, University of Twente, Control laboratory, Enschede
115. Van der Linden RR, de Groot PCJ (2007) Analyse van het grijpen met rolling-link prothesevingers. Bachelor's thesis, Delft University of Technology, Department of BioMechanical Engineering, Delft
116. Van Maanen M (2009) A simulation model of an automated item picking workstation. Internship report, Eindhoven University of Technology, Department of Mechanical Engineering, Systems Engineering Group, Eindhoven
117. Verriet J, van Wijngaarden B, van Heusden E, Hamberg R (2011) Automating the development of agent-based warehouse control systems. In: Trends in practical applications of agents and multiagent systems, Advances in intelligent and soft computing, vol 90. Springer, Berlin, pp 59–66
118. Vidal Troitinho V (2009) Design and simulation of a reconfigurable underactuated finger. Internship report, Delft University of Technology, Department of BioMechanical Engineering, Delft
119. Wassink M (2011) On compliant underactuated robotic fingers. Ph.D. thesis, University of Twente, Enschede
120. Wassink M, Carloni R, Brouwer DM, Stramigioli S (2009) Novel dexterous robotic finger concept with controlled stiffness. In: Proceedings of the 28th benelux meeting on systems and control, p 115
121. Wassink M, Carloni R, Poulakis P, Stramigioli S (2009) Digital elevation map reconstruction for port-based dynamic simulation of contacts on irregular surfaces. In: Intelligent robots and systems, 2009. IROS 2009. IEEE/RSJ international conference on, pp 5179–5184
122. Wassink M, Carloni R, Stramigioli S (2010) Compliance analysis of an under-actuated robotic finger. In: Biomedical robotics and biomechatronics

(BioRob), 2010 3rd IEEE RAS and EMBS international conference on, pp 325–330
123. Wassink M, Carloni R, Stramigioli S (2010) Port-Hamiltonian analysis of a novel robotic finger concept for minimal actuation variable impedance grasping. In: Robotics and automation (ICRA), 2010 IEEE international conference on, pp 771–776
124. Wassink M, Stramigioli S (2007) Towards a novel safety norm for domestic robotics. In: Intelligent robots and systems, 2007. IROS 2007. IEEE/RSJ international conference on, pp 3354–3359

Index

R. Hamberg and J. Verriet (eds.), *Automation in Warehouse Development*,
DOI: 10.1007/978-0-85729-968-0, © Springer-Verlag London Limited 2012

9780857299673